.en

GLOSSARY
of
RELIABILITY
and
MAINTENANCE
Terms

Gulf Publishing Company
Houston, Texas

GLOSSARY

of

RELIABILITY

and

MAINTENANCE

Terms

TED
MC KENNA

RAY
OLIVERSON

GLOSSARY *of*
RELIABILITY *and*
MAINTENANCE *Terms*

Gulf Publishing Company
Book Division
P.O. Box 2608, Houston, Texas 77252-2608

10   9   8   7   6   5   4   3   2   1

Library of Congress Cataloging-in-Publication Data

Mc Kenna, Ted.
    Glossary of reliability and maintenance terms / Ted Mc Kenna,
Ray Oliverson.
       p. cm.
    ISBN 0-88415-360-6
    1. Plant maintenance—Dictionaries.   2. Reliability
(Engineering)—Dictionaries.   I. Oliverson, Ray.   II. Title.
TS9.M43   1997
658.2′02′03—dc21   97-17097
                                                     CIP

Printed on Acid-Free Paper

*Dedicated to the memory of Ray Libby,
our friend and colleague.*

*The undiscovered country from
whose bourn no traveller returns.*

William Shakespeare, *Hamlet*

v

Acknowledgments

W e want to thank our fellow consultants Bruce Abbott, Ken Brook, William Conner, Andrew Ginder, Sharon Johnson, Richard Lowell, Charles Spillman, Greg Taylor, Glyn Thorman, and John Turner for their suggestions, contributions, and assistance in preparing this glossary. A special thanks is extended to Louis A. Bigliardi, Marathon Oil Company.

Preface

Over the nearly 25 years that we have spent assisting clients in improving maintenance and reliability, we have never failed to be puzzled at the varied words and definitions people use to denote key elements of this major area of business. A quick review follows.

Repair maintenance, responsive maintenance, routine maintenance, and reactive maintenance are terms people seem to use interchangeably to describe the day-to-day type of maintenance that is expensed by plant accountants. Also, people seem to define this activity in terms of repairing or fixing equipment or facilities. In truth, "maintain" is defined by *Webster's New World Dictionary* as *to continue in, or preserve, the same condition.* It would serve our industry well if we could settle on one term, such as expense maintenance, and forever forget the connotation of "fix." Expense maintenance involves the effort that plants expend to differentiate larger, non-routine maintenance that cannot be capitalized. Over the years, we have concluded that the effort is worthwhile in order that maintenance management can segregate tank, vessel, or reactor cleaning-type activities from real routine maintenance.

During our visits to hundreds of plants, we have heard the term *preventive maintenance,* also known as preventative maintenance, used to describe efforts ranging from all nonemergency maintenance to intervention based on time, volumes, condition monitoring, or statistical predictions based on data from equipment histories. It may be better to narrow the definition of preventive maintenance to mean intervention by maintenance or operating personnel, based on time periods or measures of usage (for example, miles driven) that is intended to mitigate the frequency or severity of equipment failure. Another issue in this area relates to the decision of whether shutdown, rebuild, or turnaround maintenance is preventive in nature or just a special type of expense maintenance. We do not believe that it should be categorized or tracked as normal preventive maintenance. To do this would inflate preventive maintenance costs and blur performance indicators.

We have encountered a great deal of confusion about the term *predictive maintenance.* Most often it is used to describe the employment of condition monitoring or nondestructive testing techniques, which vary from vibration monitoring to oil sampling, to thermography, to ultrasound. Condition monitoring is exactly what the English words describe, that is, the monitoring of the condition of the equipment. It does not become predictive maintenance until someone, or some software program such as an expert system or knowledge-based system, analyzes the data and forecasts the need for intervention by maintenance personnel. In many ways it is the "just-in-time" type of the old "don't-fix-it-until-it-breaks" maintenance philosophy. Thank goodness that no one has termed it *predictive* maintenance. Another issue we have noticed regularly is the tendency companies have to forget that empirical analysis of failure histories is a very effective part of predictive maintenance. Some companies seem to think that only data from condition monitoring can be used to predict the need for intervention.

Another term that surfaces quite often is *maintenance prevention,* sometimes known as the prevention of maintenance. This is a valid concept that is often limited to the role designers or project engineers play in building new, or modifying existing, manufacturing units. Sometimes the word *maintainability* is used.

We believe that maintenance management should initiate a failure analysis discipline to track recurring failures and to identify and resolve root causes. In this process, management will be able to track unnecessary maintenance.

Reliability is another word that is often used in plants today. "Operable" is defined by Webster as *to be in action; work.* This leads to another word, *availability,* which leads to the concept of uptime/downtime. Is availability evaluated only against schedule?

Reliable is defined as that which can be trusted; dependable. Reliability is often used to connote an umbrella concept. Maintenance, engineering, operations and support personnel all have a role in promoting reliability. Also, reliability is not restricted to maintenance issues.

The terms *reliability-centered maintenance, risk-based maintenance, optimized-reliability-centered maintenance,* and so on, lead to another area of confusion that needs clarification. Are these really different concepts, or different names for the same concepts?

There are many more examples of words and phrases that some of us use differently, especially in the area of maintenance materials management.

In our travels, we have met many managers and executives who strive to improve their departments, plants, or corporations. Many of these business executives pursue just-in-time manufacturing, ISO9000 registration, world-class recognition, or simply survival in the competitive international business environment. The test for these adventurous souls comes when they start down the path of real change, and accept the fact that they are on the journey of continuous improvement. We, as consultants, enter the picture as the agents of change and partners attempting to illuminate the path to World-Class Maintenance®.

For our clients, ineffective communication is often a core obstacle to achieving goals. This is the message in George Bernard Shaw's famous quotation, "England and America are two countries separated by the same langauge." Our experience shows that poor organizational communication stops many a journey.

We have witnessed or participated in intense discussions on issues of reliability and maintenance. Each side fervently argues their position, but they differ slightly on the issue. Then, when an apparent *impasse* is reached, both sides discover the *true* difference between their respective positions is in the definition each is using for a key word. Once a common definition is agreed upon, the sides are in full agreement. This is, of course, a waste of time and energy, yet it is a frequent occurrence in business.

From one industry to another, one organization to another, the definitions for reliability and maintenance terms vary, and therefore become an impediment to improvement. This is particularly true for critical reliability metrics such as mean time between failure or mean time to repair.

Thirty years ago, the Department of Defense prepared military standards to help provide common definitions for various terms. Over the intervening years, technology greatly expanded the number of terms needed to properly communicate in a maintenance setting. As a result, we saw the necessity to assemble this glossary to provide a common ground for all.

Throughout this dictionary we include quotations from well-known writers such as Socrates, Lincoln, and Covey. Quotations carry important messages, provide inspiration, depict a philosophy, and are a great source of humor, or should it be mirth? The struggle continues!

Ted Mc Kenna
Ray Oliverson

®*Registration mark of HSB Reliability Technologies.*

If you can keep your head when all about you
Are losing theirs and blaming it on you;
If you can trust yourself when all men doubt you,
But make allowance for their doubting too;
If you can wait and not be tired by waiting,
Or being lied about, don't deal in lies,
Or being hated, don't give away to hating

And yet don't look too good nor talk too wise;
If you can dream—and not make dreams your master;
If you can think—and not make thoughts your aim,
If you can meet with Triumph and Disaster,
And treat those two imposters just the same,

If you can make one heap of all your winnings
And risk it on one turn of pitch-and-toss,
And lose, and start again at your beginnings
And never breathe a word about your loss . . .
If you can talk with crowds and keep your virtue,
Or walk with Kings nor lose the common touch,
If neither foes nor loving friends can hurt you,
If all men count with you, but none too much;
If you can fill the unforgiving minute
With sixty seconds' worth of distance run,
Yours is the Earth and everything that's in it,
And—which is more—you'll be a Man my son!

Rudyard Kipling

A

ABC classification. Classification of a group of items in decreasing order of annual dollar volume (price multiplied by projected volume) or other criteria. Group A usually represents 10–20% of the number of items and 50–70% of the projected dollar volume. Group B represents 20% of the items and 20% of the dollar volume. Group C is 60–70% of the items and 10–30% of the dollar volume. *Synonyms:* ABC analysis, 80–20 rule, Pareto's law.

ABC inventory control. An inventory control approach based on the ABC classification.

Abscissa. In plane Cartesian coordinates, the horizontal or *x*-coordinate of a point: its distance from the *y*-axis measured parallel to the *x*-axis.

Absolute pressure. Pressure measured with respect to absolute zero pressure, as distinct from pressure measured with respect to some standard pressure; absolute zero pressure is a condition totally devoid of pressure, including the pressure of the atmosphere.

Absolute temperature. A temperature at which zero is a condition absolutely free of heat and equivalent to minus 459.72° Fahrenheit or minus 273.18° centigrade.

Accelerated deterioration. Deterioration of equipment caused by human factors that occurs over a relatively short period of time.

Acceptable performance. A measure of the amount key parameters of the parts drift beyond the acceptable limits during a specified mission life; reliability specification.

Acceptable quality limits (AQL). The percentage of defects allowed before rejecting a lot of parts, components, or product; sometimes referred to as average out-going quality limit (AOQL).

Acceptance test. A test conducted under specified conditions using delivered or deliverable items in order to determine the item's compliance with specified requirements; includes acceptance of first production units.

Access. To gain entry to part of a system or piece of equipment.

Accessibility. A measure of the relative ease of admission to the various areas of an item for the purpose of operation or maintenance.

Accountability. Answerable, but not necessarily personally charged with doing the work. Accountability cannot be delegated, but it can be shared.

Accuracy. (1) The degree of freedom from error or the degree of conformity to standard. (2) The nearness of the average of repeated measurements versus the true value (an accepted standard). *Synonyms:* deviation, error, how close the value is to being correct. (3) In alignment, ratio of the error in a measurement to the ideal or exceeded value, usually expressed in percent.

Achieved. Obtained as the result of measurement.

Active inventory. The raw materials, items, components, work in process, and finished products that will be used or sold within a given period; the group of items assigned to an operational status.

Active redundancy. That redundancy wherein all redundant items are operating simultaneously.

This is an age in which one cannot find common sense without a search warrant.

George F. Will

America's business problem is that it is entering the twenty-first century with companies designed during the nineteenth century to work well in the twentieth.

Michael Hammer and James Champy

Active time. That time during which an item is in an operational inventory.

Actual downtime hours of all operating units. Sum of actual downtime hours for all operating units.

Actual expense maintenance hours. Total company and contractor hours paid, wage positions only.

Actual expense maintenance overtime hours. Total company and contractor overtime hours paid, wage positions only.

Actual preventive maintenance tasks accomplished. Number of maintenance work orders completed in which the work order type is preventive maintenance.

Actual production run time of all operating units. Sum of actual production run time for all operating units.

When we seek to discover the best in others, we somehow bring out the best in ourselves.

William Arthur Ward

Tough-minded leaders feel, demonstrate, and express supreme self-confidence! They are comfortable with the sharing of power, authority, and beliefs.

Joe D. Batten

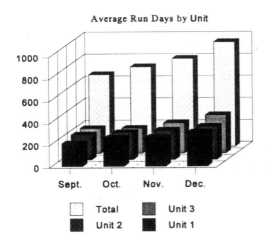

Average Run Days by Unit

Actual production volume minus off-spec product of all operating units. Sum of actual production volume minus off-spec product for all operating units.

3

Actual production volume of all operating units. Sum of actual production volume for all operating units.

Actual unscheduled maintenance downtime hours of all operating units. Sum of actual unscheduled maintenance downtime hours of all operating units.

Administrative delay. That element of delay time, not included in the supply delay time. *Synonym:* administrative time.

Affinity diagram. A total quality management tool whereby employees working in silence generate ideas and later categorize these ideas.

Air changes. The rate of air ventilation expressed as the number of times per hour that the air volume in a given room or building is changed by a HVAC (high volume air conditioning) system; equals the cubic-feet-per-hour of air flow in and out of a specified space divided by the cubic feet of the space.

Air-to-air heat exchanger. A ventilation device that warms incoming air with latent heat in exhaust air.

Air-to-air heat pump. A refrigeration or air conditioning system that removes heat from one air source and transfers it to another air stream. It is usually applied to heat inside air by cooling outside air. When operated in reverse, the system will perform like a typical air conditioning system by rejecting heat to the outside air.

AIQ. Annual issue quantity.

AISE. Association of Industrial Steel Engineers.

AITPM. American Institute for Total Productive Maintenance.

Accept that some days you're the pigeon, and some days you're the statue.

Roger C. Anderson

Thinkers prepare the revolution; bandits carry it out.

Mariano Azuela

AIV. Annual issue value.

Alert time. That element of up time during which an item is assumed to be in specified operating condition and is awaiting a command to perform its intended mission.

Algorithm. A set of rules specifying a sequence of actions taken to solve a problem.

Alignment, generic. Performing the adjustments that are necessary to return an item to specified operation.

Alignment, shaft. Positioning two or more machines so that their rotational centerlines are collinear at the coupling point operating conditions. Only the rotating shaft centerlines of different machines are aligned.

Alkaline. A substance having a pH greater than 7.0 or neutral; having alkali properties; opposite to acidic.

Allocated item. In materials requirements planning system, an item for which a picking order has been released to the stockroom, but not yet sent from the stockroom.

Allocation. The process by which a top-level quantitative requirement is distributed among lower hardware items in relation to design characteristics, reliability, and maintainability features.

AMI. Average monthly issue rate.

Amplifier. A device or instrument whose output is an increased function of an input signal, drawing power from a source other than from the input signal.

Analog. As applied to an electrical or computer system, the capability of representing data in

Yearn to understand first and to be understood second.

Beca Lewis Allen

There is no security on this earth, there is only opportunity.

General Douglas MacArthur

5

continuously varying physical phenomena (as in a voltmeter) and converting them into numbers.

Analysis. (1) A step-by-step process of determining the solution to a problem. (2) The collection and viewing of data and information, and drawing conclusions from the data and information. (3) The process of determining the composition of a substance or material using chemical or physical methods.

Analysis of variance (ANOVA). A basic statistical technique for analyzing experimental data. It subdivides the total variation of a data set into meaningful components parts associated with specific sources of variation to test a hypothesis on the parameters of the model or to estimate variance components.

You can make your world so much larger simply by acknowledging everyone else's.

Jeanne Marie Laskas

Angularity. In alignment, the angles between two centerlines.

Annunciator. A device that gives audible or visible warning or alarm when a measured process variable differs from a predetermined value.

A true measure of your worth includes all the benefits others have gained from your success.

Cullen Hightower

ANSI. American National Standards Institute.

API. American Petroleum Institute.

Appraisal costs. Costs budgeted and incurred to inspect and test the products to ascertain the level of quality and reliability attained; costs associated with measuring, evaluating, or auditing products, components, and purchased materials to ensure conformance with quality standards and performance requirements.

Apportionment. An iterative process completed in order to obtain subsystem goals that would reasonably match subsystem capabilities.

Approved equal material. Equipment or methods approved for use in a construction project as being acceptable as the equivalent in essential attributes to the material, equipment, or methods specified in the contract documents.

Area maintenance. Maintenance performed within a designated area within the plant by the maintenance shop located in the area. It often involves an area cross-functional team to evaluate, prioritize, plan, and schedule routine, extraordinary, preventive, and/ or capital work.

Arrenius law. The rate of chemical reaction is doubled (approximately) for every $10°$ centigrade rise in temperature.

The healthiest competition occurs when average people win by putting in above-average effort.

Colin Powell

Arrow diagram. A technique to determine the relationships and precedence of different activities and the time estimate for project completion. The technique is useful in identifying potential problems and improvement opportunities.

As-built drawings. Construction drawings revised to show significant changes made during the construction process.

ASCII. American Standard Code for Information Interchange.

A man of genius makes no mistakes. His errors are the portals of discovery.

James Joyce

As-found-as-left sheet. A document recording the condition of equipment prior to repair and after repair is completed. This document serves as a record of what changed due to maintenance and becomes part of the equipment's history.

ASME. American Society for Mechanical Engineers.

ASNT. American Society for Nondestructive Testing.

ASQC. American Society for Quality Control.

Asset utilization. (1) In operations, the percentage of time a plant uses to make first quality product at a 100% rate. (2) In finance, the ratio of net income to total fixed assets. Factors to consider with fixed assets include repairs and maintenance, fixed asset acquisitions to total gross assets, machinery output levels (as in #1 above), specialized assets, and equipment subject to governmental pollution requirements.

Asset Utilization
As Percent of Potential Capacity

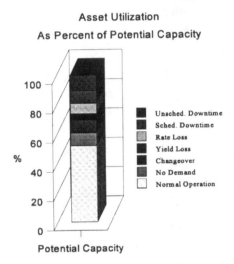

Potential Capacity

The rate at which a person can mature is directly proportional to the embarrassment he can tolerate.

Doug Engelbart

Asset value. The adjusted purchase price of the asset plus any costs necessary to prepare the asset for use.

We do have a zeal for laughter in most situations, give or take a dentist.

Joseph Heller

Assignable cause. A source of variation in a process that can be isolated, especially when its significantly larger magnitude or different origin readily distinguishes it from random causes of variation. *Synonym:* special cause.

ASTM. American Society of Testing and Materials.

Atmospheric pressure. The pressure of air at sea level; the pressure at which the mercury barometer stands at 760 millimeters or 30 inches; equivalent to 14.7 psia.

Audit. An objective comparison of actions to policies and plans.

AUP. Average unit price.

Automatic controller. A device or instrument that measures the value of a process variable and then makes alternation in flow of materials or energy to maintain the value of the variable within an acceptable range or limits.

Autonomous maintenance. Maintenance performed by equipment operators; core of this is deterioration prevention; also called operator performed maintenance (OPM).

Autonomous work group. A production team or interdepartmental team that operates a highly focused segment of the production process to an externally imposed schedule but with little external reporting, supervision, interference, or help. *Synonym:* self-directed work group (SDWT).

Availability. (1) A measure of the degree to which an item or system is in an operable and committable state at the start of a mission when the mission is called for an unknown (random) time. (The state of the equipment at the beginning of the operation includes the combined effects of the readiness-related system reliability and maintainability parameters, but excludes mission time.) (2) A measure of the operating rate; ratio of the loading time to the net operating time; calculated by subtracting downtime from loading time and dividing the product (net operating time) by the loading time; a component of overall equipment effectiveness (OEE). (3) Alternately, calculated by dividing uptime by uptime plus downtime. (4) In the process industry, calculated by subtracting from calendar time and shutdown time plus major stoppage loss and dividing the product by calendar time. Sometimes measured or expressed by dividing the mean time between failure by mean time between failure

The action and behavior of top management instills attitudes and beliefs in the workforce.

Tom Peters and Nancy Austin

The wicked leader is he who the people despise. The good leader is he who the people revere. The great leader is he who the people say, "We did it ourselves."

Lao Tsu

plus mean time to repair. Sometimes referred to as uptime or machine utilization.

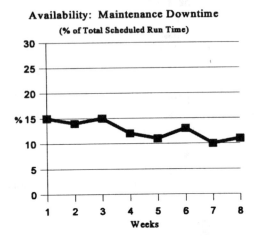

Availability: Maintenance Downtime
(% of Total Scheduled Run Time)

All business proceeds on beliefs, on judgements of probabilities, and not on certainties.

Charles William Eliot

Availability, achieved. The probability that a system or equivalent, when used under stated conditions in an ideal support environment (i.e., available tools, spares, personnel, data, etc.), shall operate satisfactorily at a given point in time. It excludes logistics time and waiting or administrative downtime. It includes active preventive and corrective maintenance downtime.

It is the nature of man as he grows older to protest against change, particularly change for the better.

John Steinbeck

Availability, inherent. The probability that a system or equipment when used under stated conditions without consideration for any scheduled or preventive action in an ideal support environment, (i.e., available tools, spares, personnel, data, etc.) shall operate satisfactorily at a given point in time. It excludes ready time, preventive maintenance downtime, logistics time, and waiting or administrative downtime.

Availability, operational. The probability that a system or equipment when used under stated conditions in an actual operational environment shall operate satisfactorily at a given point in time. It includes ready time, logistics time, and waiting or administrative downtime.

Average actual production rate. In the process industry, calculated by dividing actual production rate by operating time.

Avoidable cost. A cost associated with an activity that would not be incurred if the activity was not performed.

B

Backfill. Material used in refilling a ditch or other excavation; the process of such refilling.

Backhoe. Shovel that digs by pulling a boom-and-stick mounted bucket toward itself.

Backlog. (1) Expressed in labor-hours and weeks, represents the approved maintenance and construction work load to be done. (2) All the customer orders received but not yet shipped.

Maintenance Backlog
(Man-Hours)

Back pressure. (1) The pressure on the outlet or downstream side of a flowing system. (2) In an engine, the pressure in the exhaust system caused by obstruction that acts adversely against the piston, causing loss of power.

Baffle. Normally in the form of a plate, a partial restriction located so as to change the

Give it a chance, Ralph. Have you heard the new motto? "Influence without authority."

Sally Forth (comics)

What you do speaks so loudly that I cannot hear what you say.

Ralph Waldo Emerson

direction, guide the flow, or promote mixing within the equipment in which it is installed.

Baseline measures. A set of measurements (or metrics) that seeks to establish the current or starting level of performance of a process, function, product, firm, etc. Baseline measures are usually established before implementing improvement activities and programs.

Bathtub curve. Type of failure rate curve or pattern of failure for equipment in which the beginning or infant mortality portion shows a rapid decline to low steady or constant rate, and, after a period of time, failures again rapidly increase due to wear out.

Minor surgery is what someone else is having.

J. Carl Cook

BBL. Barrel; unit of measurement for output used in petroleum industry.

BC. Buying cost.

Don't mistake personality for character.

Wilma Askinas

Benchmarking. Process of consistently researching for new ideas for methods, practices, and processes, and of either adopting the practices or adapting the good features and implementing them to obtain the best of the best; the search for industry best practices that lead to superior performance; the continuous process of measuring one's products, services, and practices against the toughest competitors or those companies recognized as industry leaders.

Benchmark measures. A set of measurements (or metrics) used to establish goals for improvements in processes, functions, products, etc. Benchmark measures are often derived from other firms that display "best of class" achievement.

Ranges for Key Industry Measures

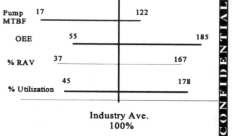

Industry Ave.
100%

CONFIDENTIAL

Benefit/cost ratio. The return on investment (ROI) multiplied by the expected years of useful life.

Best practices. Internal or external to the company or industry, those methods, processes, or approaches that represent the best way to accomplish work; generally look to the leaders in the field who demonstrate success in the area of interest; the methods used in work processes whose outputs best meet customer requirements. It is important to stress that when looking for best practices, not to limit yourself to your industry.

Bill of materials (BOM). A listing of all the subassemblies, intermediates, parts, and raw material that go into a parent assembly showing the quantity of each required to make an assembly.

Blind. A flat metal plate made to fit between piping flanges to form a positive closure of the piping.

Block diagram. A diagram that shows the operations, interrelationships, and interdependencies of components in a system.

Blueprint. In engineering, a line drawing showing the physical characteristics of a part, equipment, or system.

Boiler useful life studies (BULS). Remote field eddy current testing technology to deter-

When you cannot measure it, when you cannot express it in numbers, your knowledge is of a meager and unsatisfactorily kind.

Lord Kelvin

You can observe a lot by just watching.

Yogi Berra

mine the useful life of boilers; a proprietary process of Hartford Steam Boiler Inspection and Insurance Co.

BPO. Blanket purchase order.

Breakdown maintenance (BM). Maintenance performed after equipment fails; ideally used when failure does not significantly affect operation or production or generate any financial losses other than repair costs; remedial maintenance that occurs when equipment fails and must be repaired on an emergency or priority basis. *See* reactive maintenance.

Having patience for the process, but with a sense of urgency for the task.

Neal Davis

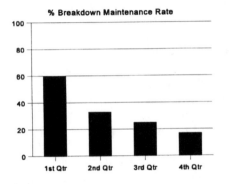

% Breakdown Maintenance Rate

Breakdown maintenance rate. Number of BM jobs divided by total maintenance jobs, expressed as a percentage.

Breakdown man-hour rate. BM man-hours divided by total maintenance man-hours.

Break-even analysis. Analysis used to determine the sales volume at which a company is able to cover all its costs without making or losing money; also known as cost-volume-profit analysis.

Break-in schedule. Time required to break-in or wear-in new or rebuilt parts, components, or equipment during the start-up phase after repair or installation.

Employees will not make sacrifices, even if they are unhappy with the status quo, unless they believe that useful change is possible. Without credible communication, and a lot of it, the hearts and minds of the troops are never captured.

John P. Kotter

British thermal unit (Btu). A unit of measure for energy or work; the heat required to raise the temperature of 1 pound of water 1° Fahrenheit.

Budget. A plan that quantifies the company's goals in terms of specific financial and operating objectives.

Budgeted capacity. The volume/mix of throughput on which financial budgets were set, and overhead/burden adsorption rates are established.

Built-in test (BIT). The self-test hardware and software that is internal to a unit to test the unit.

Built-in test equipment (BITE). A unit that is part of a system and is used for the express purpose of testing the system; BITE is an identifiable unit of a system.

Anything less than a conscious commitment to the important is an unconscious commitment to the unimportant.

Stephen R. Covey

BULS. Boiler useful life studies.

Bundle. The removable internal group of metal tubes inside a shell and tube heat exchanger.

Burn-in (preconditioning). The operation of an item under stress to stabilize its characteristics.

Progress is a nice word. But change is its motivator, and change has its enemies.

Robert F. Kennedy

Business process. A set of logically related tasks or activities performed to achieve a defined business outcome.

Business reengineering. The fundamental rethinking and radical redesign of business processes to achieve dramatic improvements (quantum, not incremental) in critical, contemporary measures of performance, such as cost, quality, service, and speed.

C

CAD/CAM. The integration of computer-aided design and computer-aided manufacturing to achieve automation for design through manufacturing.

Calendar time. The number of hours on the calendar.

Calibration. The comparison of the indication of a measuring or testing device with a known standard, the known standard itself being compared to more accurate standards in a series of controlled echelons up to a national standard held by the National Bureau of Standards, now called National Institute for Standards and Technology.

Capability index (Cpk). Measure of process capability taking into account the lack of centering of the process.

Capacity. The capability of a piece of equipment (e.g., pump) or system to perform its expected function.

Capital asset. A physical object that is held by an organization for its production potential and that costs more than some threshold value.

Capital budgeting. Actions relating to the planning and financing of capital outlays for such purposes as the purchase of new equipment, the introduction of new product lines, and the modernization of plant facilities; selection technique used to evaluate long-term investment proposals.

Capital lease. A lease transaction that is financially reported by the lessee in the same manner as if the asset acquisition had been purchased and paid for with borrowed funds; in a capital lease, the liability and the asset are both recognized on the lessee's balance sheet and reported expenses consist of interest and depreciation.

Most change programs don't work because they are guided by a theory of change that is fundamentally flawed. The common belief is that the place to begin is with the knowledge and attitudes of individuals. Changes in attitudes, the theory goes, lead to changes in individual behavior.

Michael Beer, Russell A. Eisenstat, and Bert Spector

There is nothing more difficult to take in hand, more perilous to conduct, nor more uncertain in its success, than to take the lead in the introduction of a new order of things.

Niccolo Machiavelli

Capital maintenance. Expenditures for the purchase and expansion of plant assets, generally, includes the cost of installation.

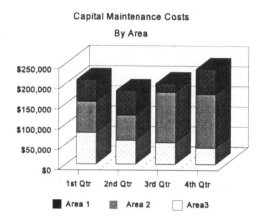

Capital Maintenance Costs
By Area

The act of asking for help may prevent us from going outside the group. It may be seen as admitting our ignorance. Be humble, we are ignorant by our very nature—admit it, and move on to creative and innovative solutions by asking for help.

Dean L. Gano

Capital projects. New construction, major repair, or improvements where the cost is capitalized (the asset is then depreciated) rather than expensed. Such projects increase the total asset value of the organization. A capital project is planned (identifying resource requirements and preparing a timetable) usually in significant detail prior to beginning activity. A return on investment (ROI) or similar calculation is performed to justify the project, and the money required for the project is budgeted for during the annual budgeting process.

If everybody is thinking alike, then somebody isn't thinking.

General George Patton

Cash flow. Operating cash flows are those that arise from normal operations and are the difference between sales revenues and cash expenses, including taxes paid. Other cash flows include issuance of stock from borrowing or from the sale of fixed assets.

Catastrophic failure. A failure that can cause complete loss of function of a component, piece of equipment, system, or unit; usually requires major maintenance, overhaul, rebuilding, or replacement.

Causal factors. All factors that might conceivably affect results, including those that are logically proved to produce the phenomena.

Cause-and-effect diagram. A tool for analyzing process dispersion; illustrates the main causes and subcauses leading to an effect. *Synonyms:* fishbone diagram, Ishikawa diagram.

Causes. Those causal factors that are proven or deduced to produce the phenomena, directly or indirectly.

Caustic soda. Sodium hydroxide solution.

Caustic solution. A corrosive, alkali mixture of caustic soda or sodium hydroxide.

c Chart. A control chart for evaluating the stability of a process in terms of the count of events of a given classification occurring in a sample; the average number of defects within each product for sample subgroups of equal sizes.

CE mark. The European Community (EC) certification symbol of conformity assessment to the specific requirements of a directive; indicates compliance with EC safety, health, and consumer legal and regulatory requirements.

Central maintenance dispatch. The selecting and sequencing of jobs to be performed by a central control group or scheduler, and the assignment of the work to craft persons, area maintenance shop, or team.

CFR. Constant failure rate.

Chargeable. Within the responsibility of a given organizational entity.

Chart of accounts. A listing of ledger account titles and account numbers used by the business.

The leader is the evangelist for the dream.

Dave Patterson

Great works are performed, not by strength, but by perseverance.

Samuel Johnson

Checklist. A tool used to ensure that important steps or actions in an operation have been taken.

Checkout. Tests or observations of an item to determine its conditions or status.

Checkout time. That element of maintenance time during which the performance of an item is verified to be a specified condition.

Check valve. An automatic valve that permits fluids to pass in one direction but closes when the fluids attempt to pass in the opposite direction.

Some go through a forest and see no firewood.

English proverb

Chemical monitoring. Monitoring that detects potential failures that cause traceable quantities of chemical elements to be released into the environment.

CIRM. Certified in integrated resource management by the American Production and Inventory Control Society (APICS).

Common sense is the least common of all the senses.

Mark Twain

CMA. Chemical Manufacturer Association.

CMI. Certified mechanical inspector by the American Society for Quality Control (ASQC).

Coaching. The managerial activity of creating, by communication only, the climate, environment, and context that empowers individuals and teams to generate results.

Coating. A material that is applied in a liquid or gel state and allowed to cure to a solid protective finish.

Collinear. Two lines that are positioned as if they were one line; in alignment, two or more centerlines or rotation with no offset or angularity between them.

Common causes. Causes of variation that are inherent in a process over item; they affect every outcome of the process and everyone working in the process. *Synonym:* random causes.

Common mode failure. Failure that has the capability to bridge and defeat the redundancy factor, causing system failure by simultaneously or sequentially impacting all redundant elements.

Company culture. A system of values, beliefs, and behaviors inherent in a company. To optimize business performance, top management must define and create the necessary culture.

Many people lose their tempers merely from seeing you keep yours.

Frank Moore Colby

Compatibility. The ability for two devices to communicate together or software to run on a particular platform.

COMPNO. Major component number.

Component. An assembly or any combination of parts, subassemblies, and assemblies mounted together in manufacture, assembly, maintenance, or rebuild; a distinct and usually replaceable part of an item.

All organizations are perfectly aligned to get the results they get.

Stephen Covey

Compounding. Situation in which some unknown or unrecognized influence is causing an effect; important factor in root cause analysis.

Computer-aided design (CAD). The use of computers in interactive engineering drawing and storage of designs; programs complete the layout, geometric transformation, projections, rotations, magnifications, and interval (cross-section) views of a part and its relationship with other parts.

Computer-aided engineering (CAE). The process of generating and testing engineering specifications on a computer workstation.

Computer-aided inspection and test (CAIT). The use of computer technology in the inspection and testing of manufacturing products.

Computer-aided manufacturing (CAM). Use of computers to program, direct, and control production equipment in the fabrication of manufactured items.

Computer-aided process planning (CAPP). A method of process planning in which a computer system assists in the development of manufacturing process plans.

Vision without action is merely a dream. Action without vision just passes time. Vision with action can change the world.

Joel Arthur Barker

Computer-aided manufacturing (CIM). The integration of the total manufacturing organization through the use of computer systems and management philosophies that improve the organization's effectiveness; the application of a computer to bridge various computerized systems and connect them into a coherent integrated whole.

All is flux, nothing stays still . . . Nothing endures but change.

Heraclitus

Computer-assisted software engineering (CASE). The use of computerized tools to assist in the process of designing, developing, and maintaining software products and systems.

Computerized maintenance management system (CMMS). Hardware and software system used to track work orders, equipment histories, and preventive/predictive maintenance schedules; usually integrated with support systems such as inventory control, purchasing, accounting, and manufacturing; computerized system to track, monitor, measure, and control maintenance and warehouse activities.

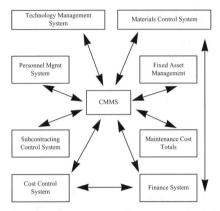

Technology Management System	Materials Control System
Personnel Mgmt System	Fixed Asset Management
	CMMS
Subcontracting Control System	Maintenance Cost Totals
Cost Control System	Finance System

Concurrent engineering (CE). A methodology for the design, development, and manufacture of products that meet the market/customer demand for high quality, low cost, and fast delivery; the application of Total Quality Management philosophy, tools, and techniques to the design-manufacturing process in combination with the tools and techniques of computer aided design (CAD) and manufacturing (CAM). *Synonym:* participative design/engineering.

Condition-based maintenance (CBM). A set of maintenance analysis techniques developed for individual equipment items, such that equipment is performed based solely on the condition of that equipment. For stationary equipment, CBM is typically driven by PM and PdM inspections; for rotating equipment, performance tests often are used to set CBM intervals.

Condition monitoring. (1) A technique used to determine when a family of devices is beginning to incur failures that need attention. (2) Equipment is used to monitor the condition of other equipment while the equipment is operating.

Condition monitoring maintenance. The establishment of a maintenance requirement by observing that a component has failed, or the detection of an impending failure by the opera-

Whatever is worth doing at all, is worth doing well.

Philip Dormer Stanhope

The leader who expects his people to perform their best will achieve the greatest results.

Joe D. Batten

tor through route monitoring during normal operation.

Conductivity. A measure of the ability of a material to conduct electricity.

Configuration audit. A review of the product against the engineering specifications to determine whether the engineering documentation is accurate, up-to-date, and representative of the components, subsystems, or systems being produced.

Conformance. An affirmative indication or judgment that a product or service has met the requirements of a relevant specification, contract, or regulation.

Whether called "taskforces," "quality circles," "problem-solving groups," or "shared-responsibility teams," such vehicles for greater participation at all levels are an important part of an innovating company.

Rosabeth Moss Kanter

Conformance testing. The running of a standard set of tests to determine whether a product meets a set of standards and/or specifications.

Consigned stock. Inventories, generally of finished product, that are in the possession of customers, dealers, agents, etc., but remain the property of the manufacturer by agreement with those in possession.

Constant failure rate (CFR). Once the failure due to components and workmanship is eliminated for the most part, the constant failure rate period is entered; also called the random failure period. The constant failure rate period is the most common time frame for making reliability predictions and is where the exponential distribution is used.

That small spark of desire to change typically comes from a key person who cares sufficiently about the organization to invest the time, energy and risk in taking leadership.

William Ouchi

Constant stage. Long stable stage of failure rate occurring after the early life or infant mortality stage and the wear-out stage; failures are of a random or chance variety. *Synonym:* chance stage.

Constant-volume heat system. An air-conditioning system that maintains space temperature conditions based on a fixed flow of air to each space or zone. Temperature variations are provided by a reheat device (for example, hot water or an electric coil) that collects thermal energy to raise the temperature of the air, thereby reducing its cooling capacity. If reheated above the space temperature, supply air can be used to heat the zone. This type of system is accurate for temperature control, but it is energy intensive.

What gets measured gets done.

Mason Haire

Constraint. Any element or factor that prevents a system from achieving a higher level of performance with respect to its goal. Constraints can be physical, such as a machine center or lack of material, but they can also be managerial, such as a policy or procedure.

Construction contractor. A licensed firm, company, or individual contracting to perform construction services in the form of new facilities, modifications, or additions to existing facilities.

Sometimes when I consider what tremendous consequences come from little things . . . I am tempted to think . . . there are no little things.

Bruce Barton

Construction cost. The cost of all of the construction portions of a project, generally based upon the sum of the construction contract or contracts and other direct construction costs.

Construction document. Document prepared when final calculations, working drawings, specifications and detailed cost estimates are completed and required drawing approvals are obtained.

Construction loads. Provisions provided to ensure that stresses due to wind loads and dead loads, and from material storage and erection equipment occurring during the erection of any structure do not exceed the allowable stresses for materials as limited by the provisions of the applicable code.

Contained drain system. A segregated, closed drain system for the removal and subsequent collection of biological or other waste that will eventually either be chemically and/or physically decontaminated before it is introduced into the sewer system or disposed of per federal, state, or local codes.

Continuous improvement (CI). Improving the quality of every process; reducing the variation in output of a process; driving defects and errors to zero; completing tasks in a shorter time; and, in the case of industrial processes, driving process variation toward zero.

Continuous process improvement (CPI). A never-ending effort to expose and eliminate root causes of problems; small-step improvement as opposed to big-step improvement.

Continuous Process Improvement

```
            ┌──────────────┐
     ──────▶│   Process    │──────▶
        ▲   │   System     │   │
        │   └──────────────┘   │
        │          │           │
        │          ▼           │
        │   ┌──────────────┐   │
        │   │   Measure    │◀──┘
        │   └──────────────┘
        │          │
        │          ▼
        │   ┌──────────────┐
        └───│   Analyze    │
            └──────────────┘
```

This is the true joy in life—that being used for a purpose recognized by yourself as a mighty one. That being a force of nature, instead of a feverish, selfish little clod of ailments and grievances complaining that the world will not devote itself to making you happy.

George Bernard Shaw

Contributing cause. The cause that contributed to the primary effect but, by itself, would not have caused the problem.

Controllable costs. Cost of those activities that are planned and included in the quality and reliability budget; includes all the activities to assure quality and reliability and all the activities to inspect and test to find out what quality and reliability level has been accomplished.

Controlled variable. The quantity or condition that is being measured and regulated by an instrument system.

Control point. The valve of a controlled variable that an automatic controller operates to maintain.

Control system. A system to guide or manipulate various elements in order to achieve a prescribed result.

Control valve. A valve, usually of a globe or needle configuration, that is positioned by the output of a control system; powered by a pneumatic diaphragm, electric motor, or hydraulic cylinder.

Convection section. The portion of the furnace in which heat is transferred from hot combustion gases, through the metal tube walls, to the fluid within the tubes.

Core competencies. Bundles of skills or knowledge sets that enable a firm to provide the greatest level of value to its customers in a way that is difficult for competitors to emulate and that provides for future growth.

Corporate culture. The set of important assumptions that members of the company share; it is a system of shared values about what is important and beliefs about how the company works; these common assumptions influence the ways the company operates.

Corrective action. A documented design, process, procedure, or materials change implemented and validated to correct the cause of failure or design deficiency.

Corrective maintenance (CM). All actions performed as a result of failure to restore an item to a specified condition; includes any or all of the following steps: localization, isolation, disassembly, interchange, reassemble, align-

After a time, you may find that having is not, after all, so satisfying a thing as wanting. It is not logical, but it is often true.

Spock (*Star Trek: Amok Time*)

In the final analysis, it is not what you do for your children but what you have taught them to do for themselves that will make them successful human beings.

Ann Landers

ment, and checkout; maintenance performed to improve equipment and its components so that preventive maintenance can be carried out reliably; equipment with design weaknesses must be redesigned. *Synonym:* repair.

Corrosion. The gradual eating away of metallic surfaces as the result of oxidation or other chemical action.

Cost analysis. A review and an evaluation of actual or anticipated cost data.

Cost center. A location, person, or item of equipment (or group), in relation to which costs are assigned; the smallest segment of an organization for which costs are collected and formally reported.

Cost control. The application of procedures to monitor expenditures and performance against progress of projects and manufacturing operations with projected completion to measure variances from authorized budgets and allow effective action to be taken to achieve minimal costs.

Cost effectiveness. A measure of the value received (effectiveness) for the resources expended (cost).

Cost of goods manufactured. Costs that include direct labor, direct materials, and manufacturing overhead.

Cost of maintenance. Costs that include lost opportunities in uptime, rate, yield and quality as a result of non-functioning or poorly functioning equipment; also involves the costs associated with equipment-related degradation of the safety of personnel, property and the environment.

I have no respect at all for the tyrant. I don't think anyone should be humiliated or demeaned because of his errors or bad judgements. Nor should a leader allow the quality of work to become low because he's too weak to demand that it remain high.

Buck Rogers

However, manipulation is getting people to do something they are either not aware of or don't agree to. That is why it is so important to get each person to know up front what you are doing and why.

Kenneth Blanchard and Spencer Johnson

Cost of Maintenance

Quality Loss   Yield Loss
Rate Loss   Uptime Lost

Cost of quality (COQ). The cost incurred as a result of nonconformance; major sources of COQ are quality of design and quality of conformance; quality costs include preventive costs, appraisal costs, internal failure costs, and external failure costs.

Cost of reliability. The total cost a manufacturer incurs during the design, manufacture, and warranty period of a product of a given reliability; can be developed around the generally accepted notion of cost of quality.

Cost per standard hour. Calculated by multiplying actual hours times labor rate, and dividing the product by standard hours produced.

Maintenance Cost Per Standard Hour

Cost reduction. The act of lowering the cost of goods or services by securing a lower price, reducing labor costs, etc. In cost reduction, the item usually is not changed, but the circumstances around which the item is secured are changed, as opposed to value analysis, in which the item itself is actually changed to produce a lower cost.

Coverage. Ratio of actual hours worked on planned work to total hours worked less delay hours.

CPIM. Certified in production and inventory management by the American Production and Inventory Control Society (APICS).

What I need is someone who will make me do what I can.

Ralph Waldo Emerson

CPM. Critical path method.

CQA. Certified quality auditor by the American Society for Quality Control (ASQC).

CQE. Certified quality engineer by the American Society for Quality Control (ASQC).

The power of accurate observation is commonly called cynicism by those who have not got it.

George Bernard Shaw

CQM. Certified quality manager by the American Society for Quality Control (ASQC).

CQT. Certified quality technician by the American Society for Quality Control (ASQC).

CR. Corrective repair; criticality.

What is said is more important than who said it.

Anonymous

CRAM. Coordinated rotational area maintenance.

Crane. A mobile machine used for lifting and moving loads without use of a bucket.

CRE. Certified reliability engineer by the American Society for Quality Control (ASQC).

Critical failure. A failure or combination of failures that prevent an item from performing a specified mission.

Criticality. A relative measure of the consequences of failure mode and its frequency of occurrences.

Critical path method. A network planning technique for the analysis of a project's completion time used for planning and controlling the activities in a project. By showing each of these activities and their associated times, the critical path, which identifies those elements that actually constrain the total time for the project, can be determined.

Critical thinking. A rigorous and thorough approach to examining a situation in such a way that all relevant factors and information are considered, appropriately weighed as to their impact, and analyzed, leading to a logical conclusion.

Cross-crafting. The training and qualifying of a maintenance craftsperson in other craft jobs.

Cross-training. The providing of training or experience in other different jobs or positions.

Common sense is perhaps the most equally divided, but surely the most underemployed talent in the world.

Christiane Collange

Nature always favors the hidden flaw.

Murphy's Third Law

Hours, Craft Cross-Training

Cryogenic welding. A method of joining pipes and tubing (3-inch diameter or less) without heat, flame, or the use of contaminating fluxes. This welding method employs special Raychem Corp. nickel alloy fittings that are packed in a cryogenic (cold) storage container. They shrink to a high-pressure fit when exposed to any ambient temperature in the range of 65 to 575 degrees Fahrenheit.

CSQE. Certified software quality engineer by the American Society for Quality Control (ASQC).

Current cost. The current or replacement cost of labor, material, or overhead. Its computation is based on current performance or measurements, and it is used to address today's costs before production as a revision of annual standard costs.

It's all very well in practice, but it will never work in theory.

French management saying

Customer service. (1) Ability of a company to address the needs, inquiries, and requests from customers. (2) A measure of the delivery of a product or service to the customer at the time the customer specified. (3) Concept of taking care of the needs of internal customers, that is, within the company.

More people would learn from their mistakes if they weren't so busy denying that they made them.

Anonymous

Cycle counting. An inventory accuracy audit technique where inventory is counted on a cyclic schedule rather once a year. Key purpose of cycle counting is to identify items in error, thus triggering research, identification, and elimination of the cause of the errors.

Cycle stock. The most active components of the inventory.

D

Database. A collection of structured information.

DBA. Database administrator.

DCF. Discounted cash flow.

DCS. Distributed central system or distributed centralized system.

De-bugging. A process to detect and remedy inadequacies.

Decibels. A measure of power gain or loss relative to an arbitrarily chosen power level; equal to 10 times the logarithm of the ratio between output power and base-level power; as applied to speech and sound levels, dB is measurement of sound energy ratios.

Decision matrix. A matrix used by teams to evaluate problems or possible solutions, where solution rating criteria are selected and the solutions are scored to determine the best solution.

Decision tree. A method of analysis that evaluates alternative decisions in a tree-like structure to estimate values and/or probabilities. Decision trees take into account the time value of future earnings by using a rollback concept.

Decomposition. The breaking up of compounds into smaller chemical forms through the application of heat, change in other physical conditions, or introduction of other chemical bodies.

Dedicated capacity. A work center that is designated to produce a single item or a limited number of similar items; proven capacity calculated from actual performance data, usually expressed as the average number of items produced multiplied by the standard hours per item.

Degradation. A gradual impairment in ability to perform.

Delay. Ratio of delay hours to total hours worked.

We don't live in a world of reality, we live in a world of perceptions.

J. Gerald Simmons

A thought that does not result in action is nothing much, and an action that does not proceed from a thought is nothing at all.

Georges Bernanos

Delay time. That element of downtime during which no maintenance is being accomplished on the item because of either supply or administrative delay.

Delegation. Assigning a project or task to a subordinate and setting the corresponding expectations. Delegation can occur formally or informally, where the former is usually associated with performance-related issues and personnel development, and the latter occurs under impromptu circumstances, that is, when the opportunity presents itself. Delegation must be distinguished from the *process of delegation* which focuses on the follow-up activities once an assignment is made, in particular, the coaching and counseling of the employee to ensure he/she is progressing satisfactorily.

The most valuable result of all education is to make you do the thing you have to do, when it ought to be done, whether you like it or not.

Thomas Huxley

Deming circle. Concept of a continuously rotating wheel of plan-do-check-action (PDCA) used to show the need for interaction among market research, design, production, and sales to improve quality.

Demonstrated. That which has been measured by the use of objective evidence gathered under specified conditions.

Great ideas originate in the muscles.

Thomas Edison

Dendrites. Formed when water containing impurities leads to leakage paths across insulators and the transport of metal atoms resulting in short circuits.

Density. The ratio of the mass of a substance to its volume.

Dependability. A measure of the item or system operating condition at one or more points during the mission. It may be stated as the probability that an item will (a) enter or occupy any one of its required operational modes during a specified mission and (b) perform the functions associated with those operational modes.

Dependent failure. Failure which is caused by the failure of an associated item(s).

Depot maintenance. Maintenance performed at customer's major maintenance facility or at the manufacturer's facility for the item.

Depreciation. An allocation of the original value of an asset against current income to represent the declining value of the asset as a cost of that item period.

Derating. Method of selecting a part for conditions less severe than its rated conditions so as to obtain improved reliability; using an item in such a way that applied stresses are below rated values, or the lowering of the rating of an item in one stress field to allow an increase in rating in another stress field.

If you don't realize there is always somebody who knows how to do something better than you, then you don't give proper respect for others talents.

Hortense Canady

Derating factor. Calculated by dividing the maximum acceptable stress value by the rated stress value.

Derating factor test. Stress test of sample to the maximum rating of the critical parameter for the part, then acceleration factors are applied to achieve a probable failure rate that could be acceptable at derated conditions.

Adversity introduces a man to himself.

Anonymous

Design development phase. Period when the drawings have progressed sufficiently to show arrangements, major details, design features, design calculations, cost estimates, outline specifications, identification of long lead items with necessary procurement documentation, and proposed organization of the project.

Design engineering. The discipline consisting of process engineering and product engineering.

Design for manufacturing. Simplification of parts, products, and processes to improve quality and reduce manufacturing costs.

Design for reliability. Four-phase process to build reliability into a part, component, equipment, or system. The phases are concept, design and development, full scale development, and operational.

Design multiplicity. The number of individual parts or components in the design; level of product or equipment simplification.

Destructive physical analysis (DPA). The methodical dissection and inspection of unfailed parts.

Deviation. (1) The difference, usually the absolute difference, between a number and the mean of a set of numbers, or between a forecast value and the actual value. (2) The difference at any instant between the value of the controlled variable and the desired setpoint or control value.

DEWN. Distributed enterprise-wide network.

Dew point. The temperature at which vapor starts to condense.

DFR. Decreasing failure rate.

Diagnostics. The action required to identify the location of a fault to a lower level of hardware than that at which the fault was detected.

Diaphragm motor value. A control value that responds to the signal from a controller and uses air pressure as the activating force.

DIN. Do it now; a priority classification.

Direct cause. The cause which directly resulted in the occurrence.

Direct maintenance labor-hours per maintenance action (DMLH/MA). A measure of the maintainability parameter related to item

As a child, one looks for compliments. As an adult, one looks for evidence of effectiveness.

Ben Bradlee

Tell me and I'll forget. Show me, and I may not remember. Involve me, and I'll understand.

Native American saying

demand for maintenance personnel; the sum of direct maintenance labor-hours divided by the total number of maintenance actions (preventive and corrective) during a stated period of time.

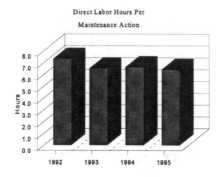

Direct Labor Hours Per
Maintenance Action

Direct work (DW). Direct maintenance to effect repair on or replace a part, component, equipment, or system; compared to indirect work that includes ordering parts, planning and scheduling the work, and completing documentation (records, permits).

Disassemble. Opening an item and removing a number of parts or subassemblies to make the item that is to be replaced accessible for removal; does not include the actual removal of the item to be replaced.

Discounted cash flow (DCF). A method of investment analysis in which future cash flows are converted, or discounted, to their value at the present time; the rate of return for an investment is that interest rate at which the present value of all related cash flow equals zero.

Dissolved inorganics. A water contaminant that includes calcium and magnesium dissolved from rock formations, gases such as carbon dioxide that ionize in water, silicates leached from sandy river beds or glass containers, ferric and ferrous ions from rusty iron pipes, chloride and fluoride ions from water treatment plants, phosphates from detergents, and nitrates from fertilizers.

No mistakes, no experience; no experience, no wisdom.

Stanley Goldstein

It's a wonderful feeling when you discover some evidence to support your beliefs.

Anonymous

Dissolved organics. A water contaminant that may include pesticides, herbicides, gasoline, and decayed plant and animal tissues; also includes the plasticizers leached out of plumbing lines, styrene monomers from fiberglass-reinforced storage tanks, deionized resin materials, and carbon from activated carbon filters.

Distillation. A water purification process involving the phased change of water from liquid to vapor and back to liquid, leaving behind certain impurities in the water phase.

Distributed control system (DCS). A real-time control system for continuous and batch process application.

The silence, often of pure innocence, persuades where speaking fails.

William Shakespeare

DNR. Do not reorder.

Documentation. The process of collecting and organization documents or the information recorded in documents.

DOP. Differential input.

Dormant. Not operating.

Dogs come when they're called; cats take a message and get back to you.

Mary Bly

DOW. Day of the week.

Downing event. The event that causes an item to become unavailable to initiate its mission; the transition from uptime to downtime.

Downtime (due to failure). Total amount of time the asset is normally out of service owing to the failure, from the moment it fails until the moment it is fully operational again; that element of active time during which an item is not in condition to perform its required function.

DP. Differential productivity.

DPWN. Distributed plant-wide network.

DRWN. Distributed refinery-wide network.

DTC. Daily time card.

DTS. Daily time sheet.

Duane plot. Predictive model measuring product reliability over time as yielded by the manufacturing process; predominantly used on electronic equipment for the measurement of the effectiveness of the reliability program.

Duct. A passageway to conduit made of sheet metal or other suitable material used for conveying air or gas at low pressures.

Duplex pump. A reciprocating, positive displacement pump having two pumping cylinders; in direct-acting steam driven pumps, having two power cylinders each connected to one of two pumping cylinders.

Durability. (1) A measure of useful life (a special case of reliability). (2) Ability to withstand use over time.

Duty. A specified operating time of an item, followed by a specified time of nonoperation. (This is often expressed as the fraction of operating time for the cycle, e.g., the duty cycle is 15 percent.)

DW. Direct work.

Dynamic joint. A joint intended to accommodate expansion and contraction movements of the structure. *Synonym:* expansion joint.

Dynamic monitoring. Monitoring that detects potential failures (especially those associated with rotating equipment) that cause abnormal amounts of energy to be emitted in the form of waves such as vibration, pulses, and acoustic effects.

Good has two meanings: it means that which is good absolutely and that which is good for somebody.

Aristotle

Blessed is the man who, having nothing to say, abstains from giving wordy evidence of the fact.

George Eliot

E

E. Effectivity.

EAC. Estimated annual consumption.

Effective. Doing the right things.

Effects. The consequences of failures; loss of one or more of the component's functions.

Efficient. Doing the right things well. Rule: First be effective.

Effluent. A liquid waste charge from a manufacturing or treatment process. The waste may be untreated, partially treated or completely treated before discharge into the environment.

Electrical ground. Conducting connection between an electrical circuit or equipment and the earth; connection to establish ground potential.

Electrical monitoring. Monitoring techniques that look for changes in resistance, conductivity, dielectric strength and potential.

Electronic data interchange (EDI). The paperless (electronic) exchange of trading documents, such as purchase orders, shipment authorization, advanced shipment notices, and invoices, using standardized document formats.

Empirical. Derived or guided from experience, experiment, or observation alone.

Employee involvement (EI). The concept of using the experience, creative energy, and intelligence of all employees by treating them with respect, keeping them informed, and including them and their ideas in decision-making processes appropriate to their areas of expertise.

Imagination is more important than knowledge.

Albert Einstein

It is later than you think.

Sundial inscription

Downtime Reduction
Due to EI

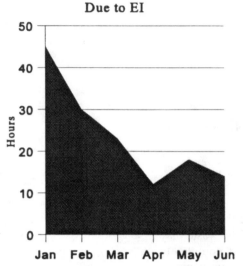

*Action is thought
tempered by illusion.*

Elbert Hubbard

Empowerment. A condition whereby employees have the authority to make decisions and take action in their work areas without prior approval.

*Nothing is easy to the
unwilling.*

Thomas Fuller

Energy units. Energy is the capability to do work; typical units of measure are Btu, joules, kilowatt-hour, foot-pounds.

Enforced problem solving. The methodology of intentionally restricting resources (e.g., inventory, storage space, number of workers) to expose a problem that must then be resolved.

Engineering change. A revision to blueprint or design released by engineering to modify or correct a part, component, equipment, or system.

Engineering standard. Design or test guidelines intended to promote the design, production, and test of part, component, or product in a manner that promotes standardization, ease of maintenance, consistency, adequacy of test procedures, versatility of design, ease of production and field service, and minimization of the number of different tools and special tools required.

Entrainment. The transport of one material in the moving stream of another material.

Environment. The aggregate of all external and internal conditions (such as humidity, temperature, radiation, magnetic and electric fields, shock vibration, etc.) either natural or manmade, or self-induced, that influences the form, performance, reliability or survival of an item.

Environmental concerns. Plant activities that impact air, water, and soil quality as driven by local, state, and federal laws (e.g., EPA) and corporate conscience; actions required to minimize or prevent and monitor adverse effects on the environment in order to comply with the laws; corrective actions taken to clean up, handle, and remove contaminated water or soil.

Environmental stress screening (ESS). Testing conducted under environmental conditions more severe that the normal ambient in which the equipment is expected to perform.

The rarest courage is the courage of thought.

Anatole France

Never go to a doctor whose office plants have died.

Erma Bombeck

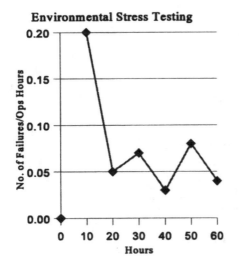

EOD. End of day.

EOM. End of month; extraordinary maintenance.

EOQ. Economic order quantity; end of quarter.

EOW. End of week.

EOY. End of year.

EPA. Environmental Protection Agency.

ER. Equipment register.

Equipment effectiveness. A measure of the value added to production through equipment.

Equipment failure loss. Time lost when equipment suddenly stops operating.

Equipment Operating Procedures (EOP). Formal written procedures for starting up, running, and shutting down equipment; critical component to operator driven reliability (ODR).

Evaluation (as related to maintainability). (1) A joint contractor and procuring activity effort to determine, at all specific levels of maintenance, the impact of the operational, maintenance, and support environment on the maintainability parameters of the item. This may include measurement of depot-level maintenance tasks. (2) The joint contractor and procuring activity effort to assess inherent maintainability or fault detection and isolation design capabilities. This would take place at the termination of validation or full-scale development phases when no specific maintainability requirements have been levied (i.e., goals or situations where just an assessment is required).

Evaporation. The conversion of a liquid into vapor, usually by means of heat.

Never tell people how to do things. Tell them what to do and they will surprise you with their ingenuity.

General George S. Patton

It is better to wear out than to rust out.

Richard Cumberland

Evaporation loss. The loss of fluid volume or weight as a result of the evaporation of the fluid.

Event maintenance. One or more maintenance actions required to effect corrective and preventive maintenance due to any type of failure or malfunction, false alarm, or scheduled maintenance plan.

Excavation. Man-made cavity or depression in the earth's surface, including its sides, walls, or faces, formed by earth removal and producing unsupported earth conditions.

Excess inventory. Any inventory in the system that exceeds the minimum amount necessary to achieve the desired throughput rate at the constraint or the quantity that exceeds the minimum amount necessary to achieve the desired due date performance.

Expansion joint. A joint or coupling designed so as to permit an endwise movement of its parts to compensate for expansion or contraction.

Expense. Items chargeable as overhead or operating costs; funds used in the maintenance, repair, or restoration of property to a sound state without extending the useful accounting life or materially increasing its value; expenditures associated with relocations and rearrangements of property.

Expense maintenance. Expenditures for maintenance and repairs necessary to the ownership and use of plant and equipment used to manufacture goods.

Expense maintenance backlog hours by craft. Total estimated labor hours on maintenance work orders awaiting execution for each craft.

Expense maintenance backlog hours by work order type. Total estimated labor hours

Getting even with somebody is not a way to get ahead of anybody.

Cullen Hightower

When you want to believe in something, you also have to believe in everything that's necessary for believing in it.

Ugo Betti

on maintenance work orders awaiting execution in each work order type.

Expense maintenance backlog hours of all operating areas. Sum of estimated labor hours on maintenance work orders awaiting execution in each operating area.

Expense maintenance input hours by craft. Total labor hours charged to maintenance work orders for each craft.

Whatever limits us we call fate.

Ralph Waldo Emerson

Expense maintenance input hours by priority. Total labor hours charged to maintenance work hours for each priority.

Expense maintenance input hours by work order type. Total labor hours charged to maintenance work orders for each work order type.

Expense maintenance input hours of all operating areas. Sum of labor hours charged to maintenance work orders in each operating area.

Our sun is one of 100 billion stars in our galaxy. Our galaxy is one of the billions of galaxies populating the universe. It would be the height of presumption to think we are the only living things within that enormous immensity.

Wernher von Braun

Maintenance Expense,
All Operating Areas

Experiment. A clearly defined procedure that results in observations.

External failure costs. Cost of unreliability during the warranty period, cost of spare parts inventories, cost of analysis, etc.; costs generated by defective products being shipped to customers.

F

Facilities. Industrial property (other than material, special tooling, special test equipment, and military property) used for production, maintenance, research, development, or tests, including real property and plant equipment.

Factor of safety. Ratio of the ultimate strength of the material to the allowable or working stress.

Fail-safe design. One in which a failure will not adversely affect the safe operation of the system, equipment, or facility.

Fail-safe device. Device whose failure on its own will become evident to the operating crew under normal circumstances; the ability of a control system to revert to a safe static condition in case of a component failure or a power failure.

Failure. The termination of the ability of a functional unit to perform its required function; loss of function when the function is needed; the event, or inoperable state, in which any item or part of an item does not, or would not, perform as specified; any event that results in work performed on equipment, rather than scheduled preventive or predictive maintenance that requires the equipment to be shut down for repair or whose lack of repair could ultimately lead to an equipment shutdown. *Synonym:* malfunction.

Failure analysis. The collection, examination, review, and classification of failures to determine trends and to identify poorly performing parts or components; subsequent to a failure, the logical systematic examination of an item,

Body: A thing shred and patched, borrowed unequally from good and bad ancestors and a misfit from the start.

Ralph Waldo Emerson

The hardest thing in the world to understand is the income tax.

Albert Einstein

its construction, application, and documentation to identify the failure mode and determine the failure mechanism and its basic cause.

Failure cause. Primary cause (or root cause) behind the failure of an item, component, equipment, or system.

Failure costs. All costs incurred as a result of failures that occur either before or after product shipment (internal and external resultant costs).

Failure diagnosis. Necessary additional quantitative tests or examinations to pinpoint the exact cause of the failure in the product performed once the failure has been verified as being a product failure.

Failure effect. The consequences a failure mode has on the operation, function, or status of an item.

Failure mechanism. Sequence of mechanical, electrical, thermal, and/or chemical processes that occurred during the period in which the failed item changed from an operational item to a failed item.

Failure mode. The effect by which a failure is observed to occur (e.g., fails to open, fails to close, fails to start).

Failure mode analysis (FMA). A procedure to determine which malfunction symptoms appear immediately before or after a failure of a critical parameter in a system. After all the possible causes are listed for each symptom, the product is designed to eliminate the problems.

Failure mode and effect analysis (FMEA). A design evaluation procedure used to identify all conceivable and potential modes and to determine the effect of each on system performance; a procedure in which each potential failure mode in every sub-item of an item is analyzed

Almost all our faults are more pardonable than the methods we resort to hide them.

La Rochefoucauld

Life is only art that we are required to practice without preparation, and without being allowed the preliminary trials, the failures and botches, that are essential for training.

Lewis Mumford

to determine its effect in other sub-items and on the required function of the item.

Failure modes, effects, and criticality analysis (FMECA). Procedure performed after a failure mode effects analysis to classify potential failure effect according to its severity and probability of occurrence; method of identifying the likely modes of failure, the possible effects of each failure and the criticality of each effect on reliability. A reliability criticality number is assigned to each failure mode based on probability of occurrence, severity of the failure effect, and the chance of being detected. These criticality numbers are used to assign priority for corrective action such as adding redundancy or devising a preventive maintenance schedule. A procedure performed after a failure mode effects analysis to classify each potential failure effect according to its severity and probability of occurrence.

Failure probability. Quantifying the chance of a component, equipment, system, or product failing within a given period of time.

Failure rate. The total number of failures within an item population divided by the total number of life units expended by that population during a particular measurement interval under stated conditions; represents the proneness of equipment failure based on age or time of operation; the number of failures over the total run hours.

Quality is never an accident; it is always the result of intelligent efforts.

John Ruskin

Take care to make things turn out well. Some people scruple more over pointing things in the right direction than over successfully reaching their goals. The disgrace of failure outweighs the diligence they showed. A winner is never asked for explanations.

Baltasar Gracián

Seal Failure Rate

Failure rate curve. A graph of the typical life of a component, equipment, system, or complex product. *Synonym:* bathtub curve.

Failure verification. First step required when test results indicate a failure has indeed resulted from something wrong with the test specimens and not from an operator error or test equipment malfunction.

Fan coil unit. A fan and heat exchanger for heating and for cooling that is assembled within a common casting. *Synonym:* fan converter unit.

FAS. Failure analysis system; in manufacturing, final assembly schedule.

Fatigue. The tendency of a metal to become brittle and fracture under conditions of repeated cyclic stressing at levels below its tensile strength.

Fault. An accidental condition that causes a functional unit to fail to perform its required function; immediate cause of failure.

Fault detection. An indication that an item is not operating within its specified operating limits.

Fault isolation. The process of determining the location of a fault to the extent necessary to effect repair.

Fault isolation time. The amount of time that is necessary to gain access to and isolate the fault (or failures).

Fault localization. The process of determining the approximate location of a fault.

Fault-tree analysis. A top-down procedure in which the analyst considers the failure modes of the overall product and works down to the identification of the causes; form of root cause analysis; a logical approach to identify the

To keep an organization young and fit, don't hire anyone until everybody's so overworked they'll be glad to see the newcomer no matter where he sits.

Robert Townsend

Perseverance is more prevailing than violence; and many things which cannot be overcome when they are together, yield themselves up when taken little by little.

Plutarch

probabilities and frequencies of events in a system that are most critical to uninterrupted and safe cooperation. This analysis may include failure mode effects analysis (determining the result of component failure interactions toward system safety) and techniques for human error prediction.

Feedback signal. A signal responsive to the value of the controlled variable in an automatic controller. The signal is returned as an input of the control system and compared with the reference signal, obtaining an actuating signal that returns the controlled variable to the desired value.

What one has to do usually can be done.

Eleanor Roosevelt

Feedwater. (1) The water entering a purification system, that is, the water brought to a filtering station before it is filtered. (2) The water supplied to a boiler to make up for evaporation loss.

Ferrography. Particle monitoring technique to detect wear, corrosion and fatigue. Wear particles are separated magnetically from lubricating oils onto an inclined glass slide by means of an instrument known as ferrograph. After separation, the total density of particles and the ratio of large to small particles indicate the type and extent of wear. The analysis is done by a technique known as bichromatic microscopic examination.

Success is relative; it is what we can make of the mess we have made of things.

T. S. Elliott

Fiber optics. A method of transmitting data and communications in which light transmitted through thin glass fibers is intensity modulated by data signals (voice or data). Normally, the use of fiber significantly reduces the large volume of wire or conductors (coaxial or twisted pair).

Fiber-reinforced concrete (FRC). Concrete-based construction material used in applications such as slabs and overlays, precast products, structural beams and girders, and shotcrete applications. Glass, polyethylene, or steel fibers are added to the other concrete ingredients. Additions of the fibers results in an improvement in strength, shock resistance, and ductility.

Fidelity. The degree to which a system accurately reproduces as its output the essential characteristics of the input signal.

Filter. (1) A porous material on which solid particles are largely caught and retained when a mixture of liquids and solids is passed through it. (2) A device for eliminating or reducing certain waves (sound, electromagnetic or optical) while leaving others relatively unchanged.

There is nothing which we receive with so much reluctance as advice.

Joseph Addison

FIN. Functional identification number.

Fire resistance. The quality of being so resistant to fire that a material or assembly, for specified time and under conditions of standard heat intensity, will not fail structurally and the side away from the fire will not become hotter than a specified temperature. Fire resistance is determined by the Standard Methods of Fire Tests of Building Construction and Materials (ASTME 119).

I've learned that you can't expect your children to listen to your words and ignore your example.

51-year-old's discovery

First in, first out (FIFO). A method of inventory valuation for accounting purposes. The assumption is that the oldest inventory (first in) is the first to be used (first out), but there is no necessary relationship with the actual physical movement of specific items.

Fishbone analysis. A technique to organize the elements of a problem or situation to aid in the determination of the causes of the problem or situation. The analysis relates the effect of the environment to the several possible sources of the problem. *Synonym:* fishbone diagram.

Fishbone Diagram

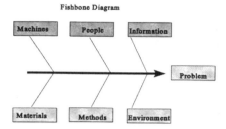

Five why's. The practice of Japanese managers to ask "why" five times when confronted with a problem. By the time they receive the answer to the fifth "why," they believe they have found the ultimate cause of the problem. *Synonym:* five W's.

Time is a dressmaker specializing in alterations.

Faith Baldwin

Fixed asset utilization. Sales divided by net fixed assets.

Fixed cost. An expenditure that does not vary with the production volume; for example, rent, property tax, and salaries of certain personnel.

There is nothing wrong with having nothing to say—unless you insist on saying it.

Anonymous

Fixed effectiveness model. Predictive model using measurements from system reliability tests. Its main purpose is to predict the adequacy (effectiveness) of individual corrective actions; used to determine when product reliability requirements will be achieved.

Flammable. Capable of being easily set on fire; combustible.

Flexible workforce. A workforce whose members are cross-trained and whose work rules permit assignment of individual workers to different tasks.

Flowchart. A chart that shows the operations, transportation, storage, delay, inspection, etc., related to a process. Flowcharts are drawn to better understand processes.

Flowchart

Focus group. A team is formed to address a specific issue, with a definite "life" for the team to recommend a solution(s) to the issue, and established measures of success. Issues are usually ideas, concepts, or methods employed by the organization.

FOP. Fixed order point.

FOQ. Fixed order quantity.

Force field analysis. A technique for analyzing the forces that will aid or hinder an organization in reaching an objective. An arrow pointing to an objective is drawn down the middle of a piece of paper. The factors that will aid the objective's achievement (called the driving forces) are listed on the left side of the arrow; the factors that will hinder its achievement (called the restraining forces) are listed on the right side of the arrow.

Form-fit-function. A term used to describe the process of designing a part or product to meet or exceed the performance requirements expected by customers.

FPO. Foreign purchase order.

FR. Failure report.

In all pointed sentences, some degree of accuracy must be sacrificed to conciseness.

Samuel Johnson

The test and the use of man's education is that he finds pleasure in the exercise of his mind.

Jacques Barzun

Frequency. A measure of pitch that distinguishes a high sound or note from a low one, measured in Hertz; periodic motion expressed as vibrations per unit time, i.e., cycles/second.

Friction. Resistance to the motion of one surface against another.

Fugitive emissions. Air contaminant emissions from sources other than stacks, ducts, vents or from nonpoint emission sources.

Functional failure analysis (FFA). The process used to identify and document the system elements, functions, and failure modes that are most important to the disciplines of maintenance and logistics planning, including reliability-centered maintenance.

Functional failure modes effects analysis. Top-down analysis of the system or product in which the functional requirements are first totally defined, then an assessment is made of any combination of potential events or conditions that might impair or prevent that function.

Functional failures. Inability of any physical asset to meet a desired standard of performance.

To wisdom belongs the intellectual apprehension of eternal things; to knowledge, the rational knowledge of temporal things.

St. Augustine

Anyone who says businessmen deal only in facts, not fiction, has never read old five-year projections.

Malcolm Forbes

G

Gain. (1) The increase in signal power as the result of amplification. (2) The signal modifying action of a controller that is the reciprocal of proportional action.

Gas chromatography. An analytical method of separating similar vapors by selection adsorption through a special placed column.

Gauge. A device used as a standard of measurement, i.e., an instrument for measuring, indicating, or regulating the capacity, quantity, volume, weight, dimensions, power, amount, properties, etc., of liquids, gases, solids.

Gauge repeatability and reproducibility (gauge R&R). Evaluation of gauging accuracy by determining whether the measurements taken with it are repeatable and reproducible. It is an estimate of the variation due to both factors.

When you counsel someone, you should appear to be reminding him of something he had forgotten, not of the light he was unable to see.

Baltasar Gracián

General support equipment (GSE). Equipment that has maintenance application to more than a single model or type of equipment.

Glass fiber reinforced concrete (GFRC). Concrete-based structural materials used in applications such as curtain walls that are manufactured in a modeling process where fiberglass is blown in as the concrete is poured.

Global measures. Set of measurements that refers to the overall performance of the firm. Net profit, return on investment, and cash flow are examples of financial measures. Throughput, operating expense, and inventory are examples of operational measures. Overall equipment effectiveness and percent return on asset value are examples of maintenance measures.

Integrity without knowledge is weak and useless, and knowledge without integrity is dangerous and dreadful.

Samuel Johnson

Grease. Oils thickened with various soaps or dry powders.

Ground. An electrical conductor connected to the earth, or a large conductor whose potential is zero.

H

Hardness. Referring to water, the scale-forming and lather-inhibiting qualities that water possesses when it has high concentrations of calcium and magnesium ions.

Hazardous processing materials (HPM). A solid, liquid, or gas that has a degree of hazard rating in health, flammability or reactivity of 3 or 4 as ranked by Uniform Fire Code Standard Number 79-3.

Hazardous substance. Substances likely to cause death or injury by reason of being explosive, flammable, poisonous, corrosive, oxidizing, irritating, or otherwise harmful.

Hazardous waste. Solid wastes, which, because of their quantity, concentration, or physical, chemical, or infectious characteristics, may (a) cause or contribute to an increase in mortality and serious, irreversible, or incapacitating illness, or (b) pose a present or potential health hazard when improperly treated, stored, transported, disposed of, or otherwise managed.

Why is it that the more mistakes people make, the more paranoid they become about other people's mistakes.

Robert Half

HC. Headcount; holding cost.

Hertz (Hz). A unit of measure of frequency (cycles per second).

Heuristic. A form of problem solving in which the results or rules have been determined by experience or intuition instead of by optimization.

Summer: The season of inferior sledding.

Eskimo proverb

Hidden function. A function whose failure will not become evident to the operating crew under normal circumstances if it occurs on its own.

Histogram. A graph of contiguous vertical bars representing a frequency distribution in which the groups or classes of items are marked on the x axis and the number of items in each class is indicated on the y axis. The pictorial nature of the histogram lets people see patterns that are difficult to see in a simple table of numbers; graphic summary of variation (dispersion) in a set of data.

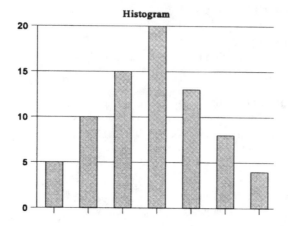

Histogram

HOD. Hour of the day.

HOOS. Hours out of service.

Hot work. Riveting, welding, burning, or other fire or spark producing operations.

Housekeeping. The manufacturing or maintenance activity of identifying and maintaining an orderly environment for preventing errors and contamination in the manufacturing process or work area.

Human factor analysis. Analysis covering safety considerations, workmanship, physical design characteristics, and maintainability considerations, and how they impact the operation of equipment and the life cycle of a product.

I

Ideal cycle time. The designed cycle time of the equipment; sometimes referred to as design speed.

IEC. International Electrotechnical Commission.

IEEE. Institute of Electrical and Electronic Engineers.

IFR. Increasing failure rate.

IIE. Institute of Industrial Engineers.

Improvement activities. Activities that extend equipment life, reduce the time required to perform maintenance, and make maintenance unnecessary.

IN. Inspector.

Inactive inventory. The group of items being held in reserve for possible future assignments to an operational status.

Inactive time. That time during which an item is in reserve.

Incremental cost. (1) Cost added in the process of finishing an item or assembling a group of items. (2) Additional cost incurred as a result of a decision.

Independent failure. Failure that occurs without being caused by the failure of any other item.

Infant mortality. (1) Failures generally due to components that do not meet specifications or workmanship that is not up to standard; not due to design related issues, but quality related issues. Failure rate declines to steady state or constant level as corrections are made. A Weibull distribution is commonly used to determine when, in fact, the mortality period is over. (2) Beginning region of failure curve showing failure rate after start-up, break-in, or wear-in period. The steep decline in this stage is caused by the "shaking out" of early-life failures. *Synonym:* early life stage.

Growing old is not for sissies.

Bette Davis

A golf course is the epitome of all that is purely transitory in the universe, a space not to dwell, but to get over as quickly as possible.

Jean Giraudoux

Inherent. Achievable under ideal conditions, generally derived by analysis, and potentially present in the design; potential.

Inherent reliability. The reliability that is commonly attributed to a good design.

Inherent reliability and maintainability value. A measure of reliability or maintainability that includes only the effects of an item design and its application, and assumes an ideal operation and support environment.

Insanity. Doing the same thing over and over again, but expecting different results.

Inspection. The qualitative observation of a component's condition or performance.

Inspection, measuring, and test equipment. All devices used to measure, test, diagnose, or otherwise examine materials, supplies, components, equipment, or systems to determine compliance with technical requirements.

Integrated resource management (IRM). A process-focused management approach (advanced by APICS) that requires cooperation throughout the process and the elimination of the traditional silo mentality or management approach. Process focus means removing functional boundaries in the product delivery system (design, production, marketing, sales, materials management, and support areas such as order entry, accounts receivable, transportation, and management information systems). The organization is managed horizontally, or across functions, to operate as an integrated whole, where the process owner understands all activities in the process and the way these activities directly and indirectly support the organization's competitive position regarding

There are a good many real mysteries in life that one cannot help smiling at, but they are smiles that make wrinkles and not dimples.

Oliver Wendell Holmes

The language of excitement is at best picturesque. You must be calm before you can utter oracles.

Henry David Thoreau

quality, time (speed of product development *and* delivery), price, and customization, as well as how it relates to making a profit now.

Interchange. Removing the item that is to be replaced and installing the replacement item.

Intermittent failure. Failure for a limited period of time, followed by the item's recovery of its ability to perform within specified limits without any remedial action.

Internal customer. The recipient (person or department) of another person's or department's output (product, service, or information) within an organization.

Internal failure costs. Yield losses caused by reliability screens and tests, cost of failure-caused manufacturing equipment downtime, cost of redesign for reliability, etc.; costs associated with defective products, components, and materials that fail to meet quality requirements and cause manufacturing losses.

Internal rate of return (IRR). Refers to the yield or interest rate that equates present value of expected cash flows from an investment project to the cost of the investment project.

International Standards Organization (ISO). A specialized international agency for standardization composed of the national bodies of 91 countries.

Interrelationship diagram. A technique used to define how factors relate to one another. Complex multivariable problems or desired outcomes can be displayed with their interrelated factors. The logical and often causal relationships between the factors can be illustrated.

Intuition. The immediate knowing or learning of something without the conscious use of rea-

In nature there are neither rewards nor punishments; there are consequences.

Robert Ingersoll

It dawned on me that it is the people on the firing line who have the power to make or break managerial decisions. Period. The people have the power whether managers want them to have it or not; whether they should have it or not. This is reality.

Ichak Adizes

soning; instantaneous apprehension; gut instinct; asset and liability in root cause analysis process.

Inventory. Those stocks or items used to support production (raw materials and work-in-process items), supporting activities (maintenance, repair, and operating supplies), and customer service (finished goods and spare parts).

Inventory control. The activities and techniques of maintaining the desired levels of items, whether raw materials, work-in-process, or finished products.

Inventory turnover. The number of times that an inventory cycles, or "turns over," during the year; frequently calculated by dividing the cost of sales by the average inventory level. *Synonyms:* inventory turns, turnover.

Illusion is the first of all pleasures.

Voltaire

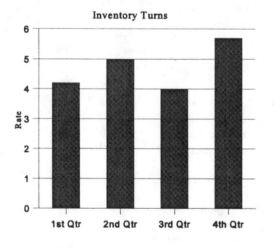

Inventory Turns

If we do not change our direction we might end up where we are headed.

Chinese proverb

IRAN. Inspection and repair as necessary.

ISA. Instrument Society of America.

ISO. International Standards Organization.

ISO 9000 series standards. A set of five individual but related international standards on quality management and quality assurance developed to help companies effectively document the quality system elements to be implemented to maintain an efficient quality system.

ISO 14001. International environmental management system (EMS) standard developed by the International Standards Organization. The standard is designed to address all facets of an organization's operations, products, and services. It covers environmental policy, resources, training, operations, emergency response, audits, measurement, and management views. It contains five major elements that an organization must satisfy to be registered or certified. These elements are policy, planning, implementation and operations, checking and corrective action, and management review.

It is better to ask some of the questions than know all the answers.

James Thurber

Issue. The physical movement of items from a stocking location.

Item. A distinct part of a unit, and usually the smallest discrete piece of equipment that is considered from an operational standpoint; a nonspecific term used to denote any product including systems, materials, components, parts, subassemblies, sets, accessories, etc.

Leadership is an example! Effective leaders deliberately set an example of what they expect and want from team members.

Joe D. Batten

ITM. Issues this month.

IVC. Inventory control.

J

J-box. A term for an electrical junction box.

JIT. Just in time.

Job description. A formal statement of duties, qualifications, and responsibilities associated with a particular job.

Job walk. A formal and documented tour of the job site to familiarize bidders or contractors with field conditions.

JTD. Job to date.

K

k. Safety factor; the Weibull shape factor.

Kaizen. The Japanese term for improvement. Continuing improvement involving everyone, managers and workers, in manufacturing; relates to finding and eliminating waste in machinery, labor, and production methods.

K-cost. The annual maintenance cost per repairman.

Key performance indicators (KPI). Set of key metrics or indicators used to track critical aspects of a business; for example, key *maintenance reliability* performance indicators are overall equipment effectiveness (OEE) and percentage return on asset value (%RAV).

When one door closes another opens, but we often look so long and so regretfully upon the closed door that we do not see the ones which opened to us.

Alexander Graham Bell

L

L. Military (MIL) symbol for 2000 failures/10^6 hours.

Labor dollars per material dollars, maintenance. A comparison of the labor and materials of the routine maintenance expense.

The body of man is a machine which winds its own springs.

De La Mettrie

Expense Maintenance Labor Dollars
Per Materials Dollar

Labor performance. Standard hours produced divided by hours worked on planned work.

Labor productivity. A partial productivity measure, the rate of output of a worker or group of workers per unit of time compared to an established standard or rate of output; expressed as output per unit of time or output per labor hour.

Last in, first out (LIFO). Method of inventory valuation for accounting purposes; the assumption is made that the most recently received (last in) is the first to be used or sold (first out), but there is no necessary relationship with the actual physical movement of specific items.

Learning curve. A curve reflecting the rate of improvement in skills as more units of an item are made. This planning technique is particularly usef''l in project-oriented industries where new products are frequently phased. The basis for the learning curve calculation is that workers will be able to produce the product more quickly after they get used to making it. *Synonym:* experience curve.

LF. Learning factor.

Life cycle. All phases through which an item, product, equipment, or system passes from conception through disposition.

Life cycle cost (LCC). The total cost of a piece of equipment or system over its entire lifetime; the total of all costs generated or forecasted to be generated during the design, development, production, operation, maintenance, and support processes. Life cycle costs include direct, indirect, recurring, nonrecurring costs such as acquisition, installation, operating, maintenance, upgrades, and removal or disposal costs.

Changing values is as important a part of reengineering as changing processes.

**Michael Hammer and
James Champy**

Better to be wise by the misfortunes of others than by your own.

Aesop

Life-limited item. An item having a limited and predictable useful life, which for reliability, safety, or economic reasons is replaced on a preplanned basis.

Life profile. A time-phased description of the events and environments an item experiences from manufacture to final expenditures or removal from the operational inventory, to include one or more mission profiles.

Life unit. Measure of product or equipment life expressed in a unit time (e.g., hours) or number of cycles of operation; a measure of use duration applicable to the item (e.g., operating hours, cycles, distances, rounds fired, attempts to operate).

Even a journey of a thousand miles begins with a single step.

Chinese proverb

Linear. A relationship in which change in one related quantity produces an exactly proportional change in the other.

Linearity. In alignment, refers to the closeness of a calibration curve to a straight line; having output directly proportional to input. *Synonyms:* straightness, direct proportion.

He is the best sailor who can steer within fewest points of wind, and exact a motive power out of the greatest obstacles.

Henry David Thoreau

Line replaceable unit (LRU). A unit which is identified for removal and replacement by organizational-level maintenance personnel.

LNG. Liquefied natural gas.

Load factor. Average load carried by an engine, machine, or plant expressed as a percentage of its maximum capacity.

Loading capacity. Safe working capacity determined in accordance with the applicable code for allowable loads and working stresses.

Loading time. Net availability of equipment during a given period of time; calculated by subtracted planned or necessary downtime from the total time available for operation.

Logic tree analysis (LTA). A structured decision process to determine the applicability and effectiveness of preventive maintenance tasks based on the failure criticality classification, the type of equipment, the failure mode and failure cause.

Logistics support. The materials and services required to operate, maintain, and repair a system. Logistics support includes the identification, selection, procurement, scheduling, stocking, and distribution of spares, repair parts, facilities, support equipment, trainers, technical publications, contractor engineering and technical services, and personnel training necessary to provide the capabilities required to keep the system in functioning status.

Logistics support analysis (LSA). A formal analytical technique to identify, define, analyze, quantify, and process logistics support requirements that requires data inputs from design, reliability, system safety, and maintainability.

LPO. Local purchase order.

LTIME. Replenishment lead time.

Lubricant. refined products of crude oil or synthetic organic hydrocarbons or inorganic materials with lubricating properties; a substance, especially oil, grease, or solid of graphite, which may be interposed between the moving parts of equipment or machinery to reduce friction, preventing contact between surfaces.

Lubricate. The adding of a type of lubrication, typically grease or oil, into a compartment, surface, or onto an exposed component.

I think knowing what you cannot do is more important than knowing what you can do.

Lucille Ball

There are risks and costs to a program of action. But they are far less than the long-range risks and costs of comfortable inaction.

John F. Kennedy

65

M

M. (1) thousand (2) maintainability (3) military (MIL) symbol for 1000 failures/10^6 hours.

Macro indicator, maintenance. A measure of end results; examples are percent replacement asset value (% RAV), asset utilization, and maintenance cost per unit of output.

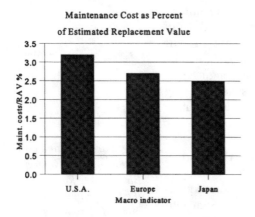

Maintenance Cost as Percent of Estimated Replacement Value

Macro indicator

Maintainability. The measure of the capability of an item to be retained in or restored to specific conditions when maintenance is performed by personnel having specified skill levels, using prescribed procedures and resources at each prescribed level of maintenance and repair; designing and installing equipment, machinery, and systems that are easy to maintain as measured by mean time to repair (MTTR); the probability that a given maintenance action for an item under given usage conditions can be performed within a time interval according to stated conditions using stated procedures and resources; maintainability has the two categories of serviceability and repairability; encompasses system testing, associated logistics, and the economics of maintenance; objective is to reduce maintenance time and cost; the characteristics of equipment design and installation that provide the ability for the equipment to be repaired easily and efficiently.

Maintainability apportionment. The assignment of maintainability subgoals to subsystems and elements thereof within a system which will result in meeting the overall maintainability goal for the system if each of these subgoals is attained.

Maintainability demonstration. The joint contractor and procuring activity effort to determine whether specified maintainability contractual requirements have been achieved.

Maintainability design criteria. A body of detailed design characteristics, which, as requirements or goals, are to be incorporated in a system or equipment design to assure that it will meet overall maintainability objectives.

Maintainability guidelines. The recommended course of action applied toward the accomplishment of the maintainability goal for a specific system or item of equipment.

Maintainability parameters. A group of factors or environmental, human, and design features that affect the performance of maintenance on equipment.

Maintainability prediction. The forecasting of quantitative maintainability characteristics of an item based on analysis of available information such as specifications, design guidelines, drawings, mock-ups, engineering models, pilot models, development equipment, maintenance environment, and experience with similar items, depending on the availability of information at the time the prediction is made.

One of the best ways to persuade others is with your ears—by listening to them.

Dean Rusk

Humor is by far the most significant activity of the human brain.

Edward De Bono

Maintainability requirement. A comprehensive statement of required maintenance characteristics (expressed in qualitative and quantitative terms) to be achieved in design and demonstrated in development.

Maintenance. All actions necessary for retaining an item in, or restoring it to, a specified condition; ensuring that physical assets continue to fulfill their intended functions.

Maintenance action. An element of a maintenance event; one or more tasks (i.e., fault localization, fault isolation, servicing and inspection) necessary to retain an item in, or restore it to a specified condition.

It is easy to be brave from a safe distance.

Aesop

Maintenance action rate. The reciprocal of the mean time between maintenance actions or 1/MTBMA (mean time between maintenance action).

Maintenance activities. Activities that prevent breakdowns and repair ailing equipment; occur in a cycle consisting of normal operation combined with preventive maintenance and corrective maintenance.

Remember that when an employee enters your office, he is in a strange land.

Erwin H. Schell

Maintenance concept. A description of the general plan for maintenance and support of an item in the operational environment. The maintenance concept provides the practical basis for design, layout, and packaging of the system and its test equipment, and establishes the scope of maintenance responsibility for each level of maintenance and the personnel resources required during normal operation.

Maintenance dollars per operating costs. Ratio of the total maintenance costs including turnarounds to total operating costs.

Maintenance dollars per RAV. Ratio of the total maintenance costs including turnarounds over the replacement asset value.

Maintenance dollars per unit of output. Ratio of the total maintenance costs including turnarounds to a standard unit of production output, for example, barrel.

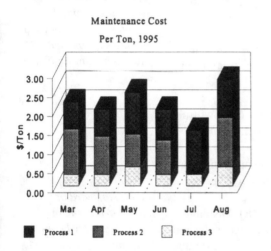

Maintenance Cost
Per Ton, 1995

Ours is a precarious language, as every writer knows, in which the merest shadow line often separates affirmation from negation, sense from nonsense, and one sex from the other.

James Thurber

Maintenance level. The three basic levels of maintenance that describe the organization or location at which maintenance is performed; these are generally called the organizational, intermediate, and depot levels.

There are an enormous number of managers who have retired on the job.

Peter F. Drucker

Maintenance prevention (MP). The reduction of maintenance costs and deterioration losses in new equipment by considering past maintenance data and the latest technology when designing for higher reliability, maintainability, operability, safety, and other requirements. *Synonym:* prevention of maintenance.

Maintenance quality assurance. The actions by which it is determined that components, equipment, or material maintained, overhauled, rebuilt, modified, or reclaimed conform to the prescribed technical requirements.

Maintenance ratio. A measure of the total maintenance manpower burden required to maintain an item. It is expressed as the cumulative number of man-hours of maintenance expended indirect labor during a given period of the life units divided by the cumulative number of end item life units during the same period.

Maintenance, repair, and operating supplies (MRO). Items used in support of general operations and maintenance such as maintenance supplies, spare parts, and consumables used in the manufacturing process and supporting operations.

Maintenance time. An element of downtime that excludes modification and delay time.

Major stoppage loss. Equipment failure loss plus production or process failure loss.

Management by example (MBE). Displaying in actions and words the characteristics desired in one's employees, that is, setting the example for your people to follow; a fundamental premise for good coaching.

Management by objectives (MBO). A participative goal-setting process that enables the manager or supervisor to construct and communicate the goals of the department to each subordinate.

Management by walking around (MBWA). Practice of physically being in touch with customers, suppliers, and/or your people listening, facilitating, teaching, coaching, and reinforcing values.

Manning level maintenance. Total authorized or assigned personnel per system at specified levels of maintenance organization.

One of the best-kept secrets in America is that people are aching to make a commitment, if they only had the freedom and environment in which to do so.

John Naisbitt

Few have heard of Fra Luca Parioli, the inventor of double-entry bookkeeping, but he has probably had more influence on human life than has Dante or Michelangelo.

Herbert J. Muller

Manufacturing check sample. A small sample of a new product, that is built by the designated manufacturing site using release engineering drawings to verify that the product can be manufactured to the drawings.

Manufacturing cycle efficiency. A measure of manufacturing efficiency calculated by dividing the processing time by the processing time plus wait time plus move time.

Manufacturing pilot run. A limited quantity run of a new product, using final processes and tooling intended for full production.

Manufacturing resource planning (MRPII). A method for the effective planning of all resources of a manufacturing company that addresses operational planning in units, financial planning in dollars, and has simulation capability to answer "what if" questions.

MAPI. Manufacturers Association of the Paper Industry.

Margin test. Qualification test concerned with assuring that the threshold of failure (the combination of conditions at which the product just begins to malfunction) is outside the range of specified conditions for the product's use.

MATCAT. Materials catalog.

Material requirements planning (MRP). A set of techniques that uses bill of material data, inventory data, and the master production schedule to calculate requirements for materials.

Material Safety Data Sheet (MSDS). Federally mandated document issued by the chemical manufacturer that describes hazards associated with each chemical product.

MAX. Maximum order = order point (OP) + order quantity (OQ).

I don't believe in just ordering people to do things. You have to sort of grab an oar and row with them.

Harold Geneen

Any third-rate engineer can design complexity.

Terry Hill

Maximum rated load. Total of all loads, including the working load, the weight of the scaffold, and such other loads as may be reasonably anticipated.

Maximum time to repair (MaxTTR). That time below which a specified percentage of all corrective maintenance tasks must be completed.

MBP. Maintenance best practices.

MBPD. Thousand barrels per day.

MDP. Maintenance department project.

Mean downtime (MDT). Average time equipment or system is down for repair.

Mean maintenance time (MMT). The measure of item maintainability taking into account maintenance policy; the sum of preventive and corrective maintenance times divided by the sum of scheduled and unscheduled maintenance events during a stated period of time.

Mean preventive maintenance. Average downtime of equipment to perform scheduled preventive maintenance.

Mean time between demands (MTBD). A measure of the system reliability parameter related to demand for logistic support; the total number of system life units divided by the total number of item demands on the supply system during a stated period of time.

Mean time between downing events (MTBDE). A measure of the system reliability parameter related to availability and readiness; the total number of system life units divided by the total number of events in which the system becomes unavailable to initiate its mission(s), during a stated period of time.

It is not the employer who pays wages—he only handles the money. It is the product that pays wages.

Henry Ford

The ability to deal with people is as purchasable a commodity as sugar or coffee. And I pay more for that ability than for any other under the sun.

John D. Rockefeller

Mean time between failure (MTBF). A basic measure of reliability for repairable items; the mean life during which all parts perform within their specified limits, during a particular measurement interval under stated conditions; an index of reliability calculated by dividing the total number of stoppages (outages) by operating time; the number of hours or cycles an item or items operated divided by the number of failures that occurred; commonly expressed as a six or 12 month rolling average; also expressed as one over the failure rate.

Mean time between maintenance (MTBM). A measure of the reliability taking into account maintenance policy; the total number of life units expended by a given time divided by the total number of maintenance events (scheduled and unscheduled) due to that item.

Mean time between maintenance actions (MTBMA). A measure of the system reliability parameter related to the demand for maintenance personnel; the total number of system life units divided by the total number of maintenance actions (preventive and corrective) during a stated period of time; the average time between maintenance actions, whether they be preventive or corrective actions.

Mean time between removals (MTBR). A measure of the system reliability parameter related to demand for logistics support; the total number of life units divided by the total number of items removed from that system during a stated period of time. This term is defined to exclude removals performed to facilitate other maintenance and removals for product improvement.

Mean time to failure (MTTF). A measure for "one shot" items or items that are thrown away after failure; the total number of operating

Don't compromise yourself. You are all you've got.

Janis Joplin

Good breeding, a union of kindness and independence.

Ralph Waldo Emerson

hours or cycles on a number of units divided by the number of failures; a basic measure of reliability for non-repairable items; the total number of life units of an item divided by the total number of failures within an item repaired at that level during a particular interval under stated conditions.

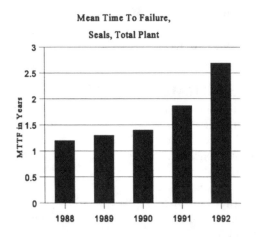

Mean Time To Failure, Seals, Total Plant

Mean time to first failure. The average time equipment can be expected to perform before it experiences its first failure; especially useful for equipment that will be installed in a use location not accessible for maintenance or repair; a reliability specification.

Mean time to repair (MTTR). An index of maintainability calculated by dividing the total stoppage time (downtime) by the operating time; the sum of corrective maintenance times at any specific level of repair divided by the total number of failures within an item repaired at that level during a particular interval under stated conditions; a basic measure of maintainability; the sum of corrective maintenance times at any specific level of repair divided by the total number of failures within an item repaired at that level, during a particular interval under stated conditions.

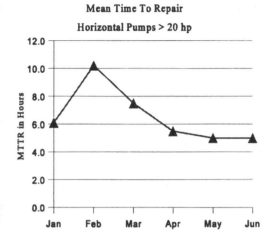

Mean Time To Repair

Horizontal Pumps > 20 hp

Mean time to restore system (MTTRS). A measure of the system maintainability parameter related to availability and readiness; the total corrective maintenance time, associated with downing events, divided by the total number of downing events during a stated period of time (excludes time for off-system maintenance and repair of detached components).

Mean time to service (MTTS). A measure of an on-system maintainability characteristics related to servicing that is calculated by dividing the total scheduled crew/operator/driver servicing time by the number of times the item was serviced.

MESC. Material and equipment standard code.

Metrics. Measures of performance, often depicted in a graphical form; examples include backlog, schedule compliance, overall equipment effectiveness, mean time between failure, and percentage return on asset value.

Micro-inch. One one-millionth of an inch; the predominant unit of measurement in the United States for line widths and vibration tolerances in the manufacture of semiconductor circuits.

Micro indicator, maintenance. Measure of process behaviors that are believed to foster end results; examples include percent planned work, percent scheduled work, size of backlog, and stockouts.

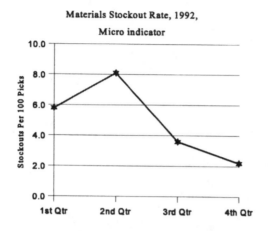

Materials Stockout Rate, 1992, Micro indicator

MIN. Minimum; the same as order point (OP).

Mission maintainability. The measure of the capability of an item to be retained in or restored to a specific condition when maintenance is performed during the course of a specified mission profile.

Mission profile. A time-phased description of the events and environments an item experiences from initiation to completion of a specified mission that includes the criteria of mission success as well as the definition of critical failures.

Mission time. That element of uptime required to perform a stated mission profile.

Mission time between critical failures (MTBCF). A measure of mission reliability; the total amount of mission time divided by the total number of critical failures during a stated series of missions.

Mission time to restore functions (MTTRF).
A measure of mission maintainability; the total corrective critical failure maintenance time divided by the total number of critical failures during the course of a specified mission profile.

Mixing value. A value that creates turbulence within a pipe to effect mixing of the materials flowing through the pipe.

MJP. Maintenance job plan.

MM. Million.

Model. A representation of a process or system that attempts to relate the most important variables in the system in such a way that analysis of the model leads to insights into the system.

Modification. upgrading of a component by changing the design, installation, or fabrication of the component.

Modification time. The time necessary to introduce any specific change(s) to an item to improve its characteristics or to add new ones.

Monitor. Observation and recording of a specified parameter that can be used to assess equipment condition; for example, vibration monitoring, thickness measurements, oil analysis.

MOP. Maintenance optimization program.

Motor valve. A valve incorporated in automatic control systems to regulate the rate of flow of material through a section of pipe; it is actuated either mechanically, electrically, or by gas pressure from a control instrument.

MPI. Maintenance performance indicator.

MR. Materials requisition.

MRD. Maximum reasonable demand.

Some people approach every problem with an open mouth.

Adlai Steveson

Nothing will ever be attempted if all possible objections must first be overcome.

Samuel Johnson

MRE. Maintenance reengineering.

MRO. Material release order.

MRP. Material requirements planning.

MRPII. Manufacturing resource planning.

MTBS. Mean time between shutdowns.

MTD. Mean time down; month to date.

MU. Maintainable unit; maintenance unit.

μ (mu) Chart. Depicts the number of errors or defects per item from samples that can vary in size.

Multicrafting. The assignment of two more workers of different crafts to a job to expedite the repair while reducing costs; for example, fitter and electrician.

Multiskilling. The use of a craftsperson trained and qualified in two or more crafts to do job to expedite repair while reducing costs; for example, insulator and carpenter.

MWC. Maintenance work control.

MWO. Maintenance work order.

MWR. Maintenance work request.

In Chinese, the word for crisis is weiji, *composed of the character* wei, *which means danger, and* ji, *which means opportunity.*

Jan Wong

The price we pay for reliability is simplicity.

Dr. Ferdinand Porsche

N

Natural deterioration. The normal wear-out of equipment that occurs in spite of proper use and maintenance.

NEMA. National Electric Manufacturers Association.

Net present value (NPV). The present value of future earnings (operating expenses have been deducted from net operating revenues) for a given number of time periods.

Net operating time. (1) Measure of stable run time minus minor stoppages and reduced speed time. (2) In the process industry, subtract performance time losses from the operating time.

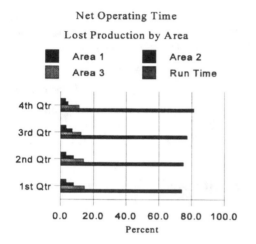

Net Operating Time

Lost Production by Area

I do not seek. I find.

Pablo Picasso

He who does anything because it is custom, makes no choice.

John Stuart Mill

Net return. Calculated by multiplying the benefit/cost ratio by the investment cost.

Neutral. A measure of pH equal to 7.0; neither acid or alkaline.

NGL. Natural gas liquids.

NIST. National Institute for Standards and Technology; formerly called the National Bureau of Standards (NBS).

Noise reduction coefficient (NRC). Refers to the absorption capability of sound absorption material used in the work environment.

Non-chargeable failure. A nonrelevant failure, or a relevant failure caused by a condition previously specified as not within the responsibility of a given organizational entity.

Nonrelevant failure. A failure verified as having been caused by a condition not present in the operational environment, or a failure verified as peculiar to an item design that will not enter the operational inventory.

Not operating. The state wherein an item is able to function but is not required to function. *Synonym:* dormant.

Not operating time. That element of uptime during which the item is not required to operate.

NPRA. National Petroleum Refiners Association.

Man must sit in chair with mouth open for very long time before roast duck fly in.

Chinese proverb

O

Occupational Safety and Health Act (OSHA). A federal law that applies to all employees in the United States who are engaged in interstate commerce. Its purpose is to ensure safe and healthful working conditions by authorizing enforcement of the standards provided under the act.

Most people are unaware that a question is almost always a stronger and more effective approach than a declarative, directive, or commanding statement.

Joe D. Batten

OH. Overhead; equipment overhaul.

Old age. Point at which a piece of equipment should be replaced because it is at the end of its useful life.

OLMICS. Off-line material inventory control system.

OM. Ordinary maintenance.

On-condition maintenance. Elimination of scheduled replacement in favor of periodic or continuous assessment to determine whether the item still functions within acceptable limits.

On-condition tasks. Tasks that entail checking equipment for potential failures so that action can be taken either to prevent the functional failure or to avoid the consequences of the functional failure.

On-stream time. The length of time a unit is in actual production. *Synonym:* utilization time.

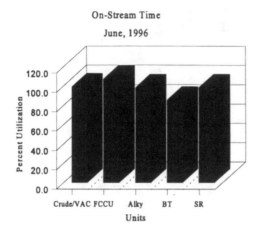

On-Stream Time
June, 1996

What another would have done as well as you, do not do it. What another would have said as well as you, do not say it. What another would have written as well, do not write it. Be faithful to that which exists nowhere but in yourself—and thus make yourself indispensable.

Andre Gidé

On-the-job training (OJT). Learning the skills and necessary related knowledge useful for the job at the place of work or possibly while at work.

OP. Order point.

Operability. Equipment or system performing to design or intended expectations; the state of being able to perform the intended function.

Operable. Capable of being put into use, operation, or practice.

Operating efficiency. A ratio (represented as a percentage) of the actual output of a piece of equipment, department, or plant as compared to the planned or standard output.

People are afraid of the future, of the unknown. If a man faces up to it, and takes the dare of the future, he can have some control over his destiny. That's an exciting idea to me, better than waiting with everybody else to see what's going to happen.

John H. Glenn, Jr.

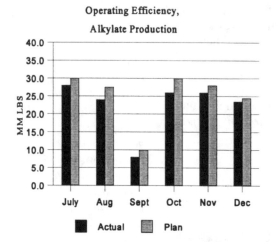

Operating Efficiency, Alkylate Production

Legend: ■ Actual   ▨ Plan

Operating leverage. Refers to the extent that fixed costs are contained in a company's cost structure; expressed as the ratio of fixed costs to total costs, percent change in operating income to the percent change in sales volume, or net income to fixed costs.

Operating time. (1) Loading time minus all major downtime. (2) In the process industry, subtract from the working time the on-stream time lost when it shuts down as a result of equipment and process failures due to operations and maintenance problems.

Operational readiness. The ability of a operations unit to respond to its operation plan(s); a function of equipment availability and status, training, etc.

Operational reliability and maintainability value. A measure of reliability or maintainability that includes the combined effect of item design, installation, quality, environment, operation, maintenance and repair.

Operator Driven Reliability (ODR). Maintenance practices performed by operators targeting reliability improvement based on preventive actions and problem analysis to optimizing equipment life cycles; consists of equipment operating procedures (EOP), operator involved maintenance (OIM), and operator performed maintenance (OPM); builds a partnership between maintenance operations/production; concept emphasizes that the operators take ownership of the equipment for which they are responsible; targets the "we operate, maintenance fixes" paradigm.

Operator Involved Maintenance (OIM). Group of activities performed by operations to address basic daily communication and coordination needs between operations and maintenance. Examples include initiating work orders, setting repair priorities, and scheduling and planning repairs, permitting, and cleanup.

Hypocrite—mouth one way, belly 'nother way.

Australian Aboriginal proverb

Operator Performed Maintenance (OPM). Minor maintenance tasks performed by operators to decrease equipment downtime and enhance product quality; principal driving activity to attain greater operator ownership of equipment and pay closer attention to equipment performance.

Being entirely honest with oneself is a good exercise.

Sigmund Freud

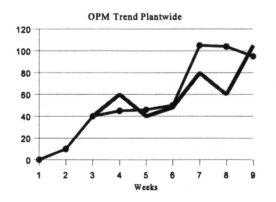

OPM Trend Plantwide

Opportunity cost. The return on capital that could have resulted had the capital been used for some purpose other than its present use.

Optimal conditions. Conditions essential for optimal functioning and maintenance of equipment capabilities.

Optimization. Achieving the best possible solution to a problem in terms of a specified objective function.

Ordinate. In plane Cartesian coordinates, the y-coordinate of a point; its distance from the x-axis measured parallel to the y-axis.

We do not imitate, but model to others.

Pericles

Organizational development. Process of improving the way in which an organization functions and is managed, particularly in response to change. Development occurs through planned intervention by a change agent in the organization's processes and is managed by upper management in accordance with the organization's overall goals.

Don't quote me; that's what you heard, not what I said.

Lawrence K. Frank

Orifice. A device for restricting the flow through a pipe. The difference in pressure on the two sides of an orifice plate can be used to measure the volume of flow through pipe.

Orifice meter. An instrument that measures the flow through a pipe by determining the difference in pressure on the upstream and downstream sides of an orifice plate.

Original equipment manufacturer (OEM). A manufacturer that buys and incorporates another supplier's products into its own products; also, products supplied to the original equipment manufacturer or sold as part of an assembly.

OSHA 1910. Set of federal regulations containing the occupational safety and health standards for industries that have been found to be national consensus standards or established federal standards.

OQ. Order quantity.

Out of spec. A term used to indicate that a unit does not meet a given specification; a piece of equipment or component not operating within expected performance parameters.

Overall equipment effectiveness (OEE). Measure of equipment performance based on its actual availability, rate of performance, and quality of product; process OEE is actual throughput divided by maximum potential throughput times 100; expressed as a percentage when multiplying availability times rate times quality.

Intelligence is quickness to apprehend as distinct from ability, which is capacity to act wisely on the thing apprehended.

Alfred N. Whitehead

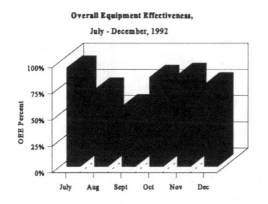

Overall Equipment Effectiveness, July - December, 1992

Experience is not what happens to a man. It is what a man does with what happens to him.

Aldous Huxley

Overhaul. Disassembly and reconditioning of a component by the removal and replacement of internal components.

Overload. A current, voltage, or power level applied to an instrument beyond which permanent damage will occur.

P

P. Productivity; military (MIL) symbol for 100 failures/10^6 hours.

PA. Price agreement.

P&ID. Piping and instrumentation diagram.

Paradigm. A set of assumptions, everyday truths, and conventional wisdom about people and how they work in organizations.

Paradigm shift. Fundamental change in the culture of an organization.

Parameter. A coefficient appearing in a mathematical expression, each value of which determines the specific form of the expression; parameters define or determine the characteristics of behavior of something.

Participative design/engineering. A term that refers to the participation of all the functional areas of the firm in the product design activity. *Synonym:* concurrent engineering.

Particle monitoring. Monitoring that detects potential failures that cause discrete particles of different sizes and shapes to be released into the environment in which the item or component is operating.

Particulate. A state of matter in which solid or liquid substances exist in the form of aggregated molecules or particles.

Part standardization. A program for planned elimination of superficial, accidental, and deliberate differences between similar parts in the interest of reducing part and supplier proliferation.

Payback method. A method of evaluating an investment opportunity that provides a mea-

In all affairs, love, religion, politics or business, it's a healthy idea, now and then, to hang a question mark on things you have taken for granted.

Bertrand Russell

When the mind is thinking, it is talking to itself.

Plato

sure of the time required to recover the initial amount invested in a project; focuses on the payback period that is defined as the amount of time a company expects to take before it recovers its initial investment.

Payback period. The reciprocal of the simple return of investment (ROI); ignores total cash flow over time and the time value of money.

Pay for knowledge. A pay restructuring scheme by which competent employees are rewarded for the knowledge they acquire before or while working for an organization, regardless of whether such knowledge is actually being used at any given time.

There is nothing wrong with mistakes. Just don't respond with encores.

Anonymous

Pay for skill. A pay restructuring scheme by which competent employees are rewarded for the skills they acquire while working for an organization, regardless of whether such skills are actually being used at any given time.

PC. Personal computer.

p chart. A control chart for evaluating the stability of a process in terms of the percentage of the total number of units in a sample in which an event of a given classification occurs over time; the proportion of nonconforming items in a sample of items. p charts are used where it is difficult or costly to make numerical measurements or where it is desired to combine multiple types of defects into one measurement.

Still round the corner there may wait,

A new road, or a secret gate.

J. R. R. Tolkien

pdf. Probability density functions.

Percent accuracy of maintenance spare parts high turnover inventory. The number of times the physical count for a line item equals the computer (record) count.

Percentage maintenance downtime. Maintenance downtime (minutes or hours) divided by the total amount of time the unit is scheduled to run.

Percentage of emergency maintenance.
Number of emergency jobs divided by the total number of jobs completed.

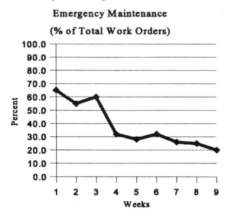

Optimism is the faith that leads to achievement.

Helen Keller

Percentage of planned maintenance. Hours spent on planned maintenance divided by total maintenance hours.

Out of intense complexities intense simplicities emerge.

Winston Churchill

Percentage of preventive maintenance work orders. A comparison of those work orders that are of a preventive maintenance nature with total work orders completed in a calendar year.

Percentage of Replacement Asset Value (% RAV). Measure of total annual maintenance investment for the plant expressed as a percentage.

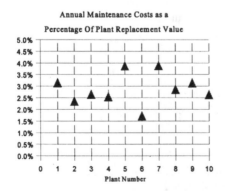

Percentage of time scheduled. Scheduled hours divided by total hours available.

Percentage overtime. Overtime hours divided by the total maintenance hours.

Percentage rework. Number of rework jobs divided by the total number of jobs.

Percentage work identification. Number of jobs performed with work orders divided by total number of jobs performed.

Percentage of work orders planned. Number of work orders planned divided by the number of work orders written.

Percent of maintenance spare parts stock-outs. Number of times stores do not have the number of spare parts that a maintenance craftsperson requests with the total number of requests for spare parts.

Performance. The degree with which an employee applied skill and effort to an operation or task as measured against an established standard.

Performance efficiency. Product of ideal cycle time times total parts run divided by the operating time; overall speed rate times net operating time; also referred to as throughput efficiency or equipment performance efficiency.

Performance measurement. Qualitative measure of cost effectiveness, functional effectiveness, productivity, quality, health, safety, management, environmental, maintenance, or energy-savings; performance indicators, success indicators.

Performance rate. (1) Product of net operating rate times the operating speed rate; ideal cycle time times output divided by operating

I'm troubled. I'm dissatisfied. I'm Irish.

Marianne Moore

It takes a very unusual mind to undertake the analysis of the obvious.

Alfred North Whitehead

time; a component of overall equipment effectiveness (OEE). (2) In the process industry, calculated by dividing the average actual production rate by standard production rate.

Performance Rate, FCCU/Gas Con

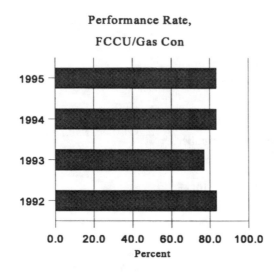

Performance standard. (1) Desired performance level, or what we want to achieve; (2) Inherent or built-in ability to perform to the equipment or systems design capability, or what it is capable of achieving.

Performance time losses. Production rate reductions due to start-up, shutdown, changeover, and abnormal production losses.

PFD. Process flow diagram.

PF(t). Probability of failure over time.

P-F interval. The interval between the occurrence of a potential failure and its decay into a functional failure; warning period or lead time to failure.

pH value. The logarithm of the reciprocal of the hydrogen (H) ion concentration; indicates the acid or alkaline condition of a substance, pure water and neutral solution having a pH of 7.0; acid solutions having a pH less than 7.0; alkaline solutions a pH greater than 7.0.

Physical monitoring. Monitoring techniques that detect potential failures in the form of cracks, fractures, the visible effects of wear, and dimensional changes.

PI. Pressure indicator.

PIMA. Paper Industry Manufacturers Association.

Piping and instrumentation diagram (P&ID). Diagram of the unit, system, or equipment showing control systems such as values and steam traps, and instruments including pressure indicators (PI), temperature indicators (TI), and level indicators (LI) or controllers.

PK. Purchasing constant.

Plan-do-check-action (PDCA). A four step process for quality improvement. *Synonyms:* Deming circle, Shewhart cycle.

Planned downtime. The amount of downtime officially scheduled in the production plan that includes downtime for scheduled maintenance and management activities (e.g., training, meetings); breaks plus lunch plus planned meetings, planned training plus preventive maintenance.

Autumn is the bite of a harvest apple.

Christina Petrowsky

An essential truth for leaders and team builders to grasp is this: We can know and lead others only when we are progressively learning how to know and lead ourselves. Self-discovery is a frequently neglected but crucial step. Remember, all growth is self-growth. Grow the example you want your team members to follow.

Joe D. Batten

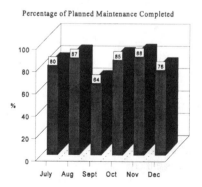

Percentage of Planned Maintenance Completed

Planned maintenance. Maintenance work with a defined scope, set sequence of work, identified craft requirements and labor-hours, estimated duration, anticipated material requirements, and prepared supporting documentation (e.g., piping diagram).

Plenum. An air compartment or chamber to which one or more air ducts are connected; part of an air handling system that supplies conditioned air, circulates air, or exhausts air.

Plume discharge velocity. The velocity of an exhaust gas from a vertical stack measured at the point of discharge. The higher the velocity of the exhaust gas, the higher the plume, and the greater the dispersion of exhaust gas into the atmosphere.

PM. Preventive maintenance; preventative maintenance; small case (p.m.) refers to the period noon to just before midnight.

P-M analysis. Technique to promote the thorough, systematic elimination of defects that contribute to chronic losses or failures.

PM chart. Measures the number of defective items in the sample.

PO. Purchase order.

POD. Plan of the day.

Polish. In water purification, the process of removing the remaining contaminants from preprocessed feedwater.

POM. Plan of the month.

Positioning action. Action in which there is a predetermined relation between the value of the controlled variable and the position of the final control element.

Potential failure. An identifiable physical condition that indicates that a functional failure is either about to occur or in the process of occurring.

We judge ourselves by our motives and others by their actions.

Dwight Morrow

To get others to come into our ways of thinking, we must go over to theirs; and it is necessary to follow, in order to lead.

William Hazlitt

Potentiometer. An instrument used to measure or compare electromotive forces; a variable resistor.

POW. Plan of the week.

Power units. The rate at which work is done; work per unit of time; units of measure including watts, joules per second, horsepower.

PPE. Personal protection equipment.

PR. Predictability ratio; priority; purchase requisition; public relations.

Precision. The lack of variation in repeated measurements to each other.

Predicted. That which is expected at some future time, postulated on analysis of past experience and tests.

Predictive maintenance (PdM). The use of modern measurement and signal-processing techniques to accurately diagnose the condition of equipment (level of deterioration) during operation; periodic measurement and trending of process or machine parameters with the aim of predicting failures before they occur; predict or anticipate when maintenance is required through condition monitoring of equipment. Examples are vibration monitoring, lubricant analysis, and leak detection. *Synonym:* predictable maintenance.

Hard work spotlights the character of people; Some turn up their sleeves, some turn up their noses, some don't turn up at all.

Sam Ewing

The secret of success is constancy to purpose.

Benjamin Disraeli

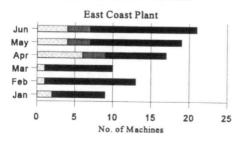

Predictive Maintenance Results

East Coast Plant

No. of Machines

☐ Repair Within The Month
▨ Immediate Attention
■ Emergency Repair
■ Machines Monitored

Pre-engineering phase. Composite, preliminary studies outlining the basic assumptions, general approach, and probable budget and timetable for the proposed facilities construction project. Such studies include process, feasibility, and preliminary engineering studies and comprise the data necessary for preparation of the appropriation request.

Present value. The value today of future cash flow.

Pressure. The force or thrust exerted on a surface, normally expressed as force per unit area. Pressure is exerted in all directions in a system.

I used to be snow-white . . . but I drifted.

Mae West

Pressure drop. The decrease in pressure due to friction, that occurs when a liquid or gas passes through a pipe, vessel, or other piece of equipment.

Pressure gauge. Device that responds to the difference between atmospheric pressure and the pressure in a closed system, such as a tank or pipeline.

Nothing fails like success because we don't learn from it. We learn only from failure.

Kenneth Boulding

Pressure vessel. An enclosed container in which a pressure greater than atmospheric pressure can be maintained.

Preventable maintenance. Work which occurred because of errors of omissions or commission on the part of project engineers, operators, mechanics, and plant management.

Preventable Maintenance

(Average Distribution)

58% Necessary
10% Maintenance
Production
Management
9%
Design/Construction
3%
20%

Preventive activities. Design for reliability, reliability standards and guidelines development, customer requirements research, product qualification, design reviews, reliability training, fault-tree analysis, failure modes, effects and criticality analysis, etc.

Preventive costs. Costs budgeted and incurred to assure acceptable quality and reliability; a controllable cost; costs caused by improvement activities that focus on the reduction of failure and appraisal costs; costs associated with personnel engaged in designing, implementing, and maintaining the quality system, including auditing system; other typical costs include education, quality training, and supplier certification.

There is no crisis to which academics will not respond with a seminar.

Old saying

I hear and I forget. I see and I remember. I do and I understand.

Chinese proverb

Preventive maintenance (PM). All actions performed in an attempt to retain an item in specified condition by providing systematic inspection and detection; any actions that are preventive of incipient failures; periodic inspection to detect conditions that might cause breakdowns, production stoppages, or detrimental loss of function combined with maintenance. The goal is to control such conditions in their **early stages** and includes basic housekeeping, periodic/systematic inspection, detection, and daily routine maintenance (e.g., adjustments, replacements) to prevent deterioration. *Synonym:* preventative maintenance.

Preventive maintenance hours worked. Total labor hours on maintenance work orders completed where work order type is preventive maintenance.

Percentage of Preventive Maintenance Hours
Compared to Responsive and Predictive

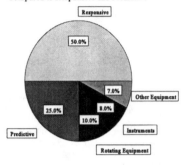

Trust me, but look to thyself.

Irish proverb

Preventive maintenance rate. Total PM jobs performed divided by total PM jobs planned.

PRI. Priority.

Price of maintenance. Maintenance budget including labor, material, outside services, and overhead.

By working faithfully eight hours a day you may eventually get to be a boss and work twelve hours a day.

Robert Frost

Priority effect. The primary effect or problem you are trying to prevent from occurring or recurring; it is singular with one or more root causes.

Proactive maintenance. The combination of operator-performed maintenance, preventive maintenance, and predictive maintenance activities whereby maintenance is conducted to prevent, eliminate, delay, or reduce maintenance before failure.

Probability. Number between 0 and 1 that indicates the likelihood of occurrence of the event; this number can be either subjective or based upon the empirical results of experimentation.

Probability distribution. The patterns represented by formulas, pictures, tables, and so on, used to portray the behavior of the random variables.

Procedure manual. A formal organization and indexing of a company's procedures.

Process. A collection of activities that takes one or more kinds of input and creates an output that is of value (whether perceived or apparent); includes any sequence of organizational activities or tasks, whether in manufacturing or the office.

Process capability (Cp). Refers to the ability of the process to produce parts or product that conform to (engineering) specifications; relates to the inherent variability of a process that is in a state of statistical control; calculated by dividing part tolerance range by probable process variation.

Process capability analysis. A procedure to estimate the parameters defining a process. The mean and standard deviation of the process are estimated and compared to the specifications, if known.

Process control. The act of or state of maintaining a process within specified limits (upper and lower limits). If a process is in control, there will be no assignable causes operating in the process, and variation seen will be due solely to random causes.

Process decision chart. A technique used to show alternate paths to achieving given goals. Applications include preparing contingency plans and maintaining project schedules.

Process discipline. Establishment or improvement of methods and procedures that enable efficiency and repeatability.

The worst crime against working people is a company which fails to operate at a profit.

Samuel Gompers

Noise proves nothing. Often a hen who has merely laid an egg cackles as if she had laid an asteroid.

Mark Twain

Process failure losses. Time lost when equipment shuts down due to factors (e.g., misoperation, shortage of raw materials) external to the equipment.

Process reengineering. A specific, finite process is identified and redesigned for improvement.

Process value analysis. A method for identifying those activities that add value to the process output and those that add cost and little or no value.

Production adjustment loss. Time lost when changes in supply and demand require unplanned adjustments in production plans due to product changeover.

Production reliability acceptance testing (PRAT). Type of testing used when assurance is needed that production deficiencies do not cause loss in reliability.

Productivity (P). An overall measure of the ability to produce a good or a service; ratio of output (goods or services) to input (cost of output); total standard hour produced divided by total actual hours working (including delays) multiplied by 100.

I am easily satisfied with the very best.

Winston Churchill

There is nothing so useless as doing efficiently that which should not be done at all.

Peter F. Drucker

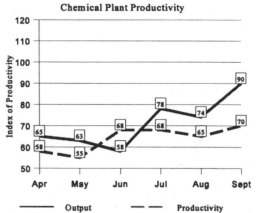

Chemical Plant Productivity

--- Output   - - Productivity

98

Program evaluation and review technique (PERT). A network planning technique for the analysis of a project's completion time. It uses an algorithm that permits identification of the critical path, the string of sequential activities that determines the project's completion time. PERT time estimates are probabilistic, based on pessimistic, most likely, and optimistic time estimates for each activity.

Project completion. Calculated by dividing estimated hours complete by the total estimated project hours and multiplying by 100.

Project management. The use of skills and knowledge in coordinating the organizing, planning, scheduling, directing, controlling, monitoring, and evaluating of prescribed activities to ensure that the stated objectives of a project, manufactured product, or service are achieved.

Be willing to have it so; acceptance of what has happened is the first step to overcoming the consequences of any misfortune.

William James

Project manual. A book containing all written contract documents except the drawings and may include the bidding documents, sample forms, general and supplement conditions.

Proportional band. The range of measured values needed to cause maximum possible change in the final control element setting, i.e., the amount of pen movement necessary to give full value movement; usually expressed as a percent of full scale range.

Character is perfectly educated will.

Novalis

Protective device. Devices designed to (1) draw operators' attention to abnormal conditions (e.g., audible alarm), (2) shut down equipment in the event of failure (e.g., fail-safe sensor), (3) eliminate or relieve abnormal conditions that follow a failure and which might otherwise cause much more serious damage (e.g., rupture disk), (4) take over from a function which has failed (e.g., redundant circuit), or (5) prevent dangerous situations from arising in the first place (e.g., security guards).

PSI. Pounds per square inch.

PSIA. Pounds per square inch, absolute.

PSIG. Pounds per square inch, gauge.

PS(t). Probability of survival over time.

p(t). The failure probability density.

PTD. Project to date.

PTNF. Predicted time to next failure.

Pump. A machine for raising or transferring liquids or gases by suction, pressure, or both.

He is a bold man that first ate an oyster.

Jonathan Swift

Punch list. The list of tasks not completed or the list of owner complaints requiring attention before a construction job is considered done.

PV. Present value.

PVC. Present value cost.

Have the courage to act instead of react.

Earlene Larson Jenks

Q

QAT. Quality action team.

QIP. Quality improvement program; quantity in-process.

QOH. Quantity on hand.

QOO. Quantity on order.

QTD. Quarter to date.

Qualification plan. A comprehensive plan detailing how a new product design or revision will be evaluated and/or tested to verify conformance to the product specification.

Qualification review panel. An independent, cross-functional body of program evaluators whose primary tasks are to approve the product qualification plans, monitor the execution of those plans, evaluate project risks, and determine the readiness of the product for production release.

Quality. Degree to which product characteristics conform to the requirements placed upon that product, including reliability, maintainability, and safety; totality of features and characteristics of a product or service that bear on its ability to satisfy given needs; fitness for use; degree of variation from the target (nominal) value; conformance to requirements.

Tradition is a guide not a jailor.

W. Somerset Maugham

Quality assurance. All those planned or systematic actions necessary to provide adequate confidence that a product or service will satisfy given needs; a planned and systematic pattern of all means and actions designed to provide adequate confidence that items or services meet contractual and jurisdictional requirements and will perform satisfactorily in service; quality assurance includes quality control.

The investigation of the meaning of words is the beginning of education.

Antisthenes

Quality defect losses. Time lost in producing rejectable product, scrap product, or downgrading of product.

Quality engineering. The analysis of a manufacturing system at all stages to maximize the quality of the process itself and the products it produces; same principles are applicable to project engineering and maintenance engineering.

Quality function deployment (QFD). Comprehensive planning process in which customer requirements deployed through interlinked requirement matrices in every activity result in a series of interlinked, interdependent, detailed descriptions of design decisions, process identification, and procedures as well as process controls.

Quality of conformance. Conformance between the product and its design specifications.

Quality of design. Conformance between customer expectations and design expectations.

Quality management. The aspect of the overall management function that determines and implements the quality policy; requires commitment and participation of all elements of the organization, where the responsibility for quality management belongs to top management; includes strategic planning, allocation of resources, and other systematic planning, operations, and evaluations.

You can never plan the future by the past.

Edmund Burke

Quality rate. (1) Calculated by quantity (number, volume) of defects from input, and divided by input; a component of overall equipment effectiveness (OEE). (2) In the process industry, calculated by subtracting from production quantity the quality defect loss plus reprocessing loss and dividing by production quantity.

Those who lose dreaming are lost.

Australian Aboriginal proverb

Losses from Rejection/Reprocessing
North American Refinery

R

r. The failure rate.

R. Reliability; military (MIL) symbol for 10 failures/10^6 hours.

Raceway. Channel for loosely holding wires or cables in interior work that is designed expressly and used solely for this purpose.

Random experiment. Type of experiment where the outcomes or observations in any one trial of such an experiment cannot be predicted with certainty.

Random failure. Failure whose occurrence is predictable only in a probabilistic or statistical sense.

Range chart. A control in which the subgroup range is used to evaluate the stability of the variability within a process; plot of the range for each sample with calculated control limits. Commonly referred to as the R chart.

Rated capacity. The expected output capability of a resource or system. Capacity is traditionally calculated from such data as planned hours, efficiency, and utilization. The rated capacity is equal to hours available times efficiency times utilization. *Synonyms:* calculated capacity, nominal capacity.

Ratio control. A type of control that uses a measurement of one variable to adjust the control point for a second variable; often used for proportioning the flow of an additive on the total flow.

Reaction time. Uptime needed to initiate a mission, measured from the time command is received.

Just because we cannot see clearly the end of the road, that is no reason for not setting out on the essential journey. On the contrary, great change dominates the world, and unless we move with change we will become its victims.

Robert F. Kennedy

Man seems to be a rickety poor sort of thing, any way you take him; a kind of British Museum of infirmities and inferiorities. He is always undergoing repairs. A machine that was as unreliable as he would have no market.

Mark Twain

Reactive maintenance (RM). Maintenance responses to equipment malfunctions or breakdowns after they occur; see breakdown maintenance.

Reactive vs. Proactive Maintenance Hours

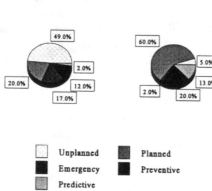

We're drowning in information and starving for knowledge.

Rutherford D. Rogers

Readout. The visual indication and printed record of any process variable.

Reassembly. Assembling the items that were removed during disassembly and closing the reassembled items.

The final test of a leader is that he leaves behind in other men the conviction and the will to carry on.

Walter Lippmann

Recipes. Control or process instructions that include: work instructions (both what to build and how to build), equipment instructions (what and how to use), operator instructions (who to use, when/how to use, safety, data collection, scheduling), machine instructions (which to use, how), and scheduling instructions (when to build).

Reciprocating pump. A positive pump consisting of a plunger or a piston moving back or forth within a cylinder. With each stroke of the plunger or piston, a definite volume of liquid is pushed out through the discharge.

Recorder. An instrument whose primary function is to make a continuous record of process conditions.

Recorder-controller. An instrument that combines the function of recording a condition with that of controlling it also.

Redesign. Any change to the specification of any item of equipment; changing the specification of a component adding a new item, replacing an entire machine with one of a different make or type, or relocating a machine; any other change to process or procedure which affects the operation of the equipment or system.

A difference of opinion is what makes horse racing and missionaries.

Will Rogers

Redundancy. The existence of more than one means for accomplishing a given function; each means of accomplishing the function need not necessarily be identical.

To have his path made clear for him is the aspiration of every human being in our beclouded and tempestuous existence.

Joseph Conrad

Regulator. (1) A device that measures a variable and limits the deviation of the value from its selected reference. (2) A specialized automatic value that reduces the pressure of a gas or liquid stream and maintains the reduced pressure over a range of low rates.

Relevant. That which can occur or recur during the operational life of an item.

Reliability. The probability that equipment, machinery, or systems will perform their required functions satisfactorily under specific conditions within a certain time period; measured by mean time between failure (MTBF); the duration or probability of failure-free performance under stated conditions.

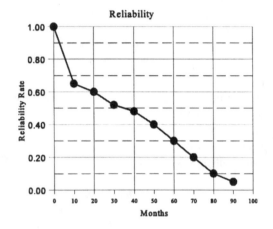

Timing, degree and conviction are the three wise men in this life.

R. I. Fitzhenry

Reliability and maintainability accounting. That set of mathematical tasks that establish and allocate quantitative reliability and maintainability (R & M) requirements, and predict and measure quantitative R & M requirements.

Each man must look to himself to teach him the meaning of life. It is not something discovered; it is something moulded.

Antoine de Saint-Exupéry

Reliability and maintainability engineering. That set of design, development, and manufacturing tasks by which reliability and maintainability are achieved.

Reliability appraisal. Life testing, environmental ruggedness evaluation, abuse testing, failure data reporting and analysis, reliability model, etc.

Reliability assurance. All actions necessary to provide adequate confidence that material conforms to established reliability requirements.

Reliability block diagram (RBD). A schematic model of a system showing series and/or parallel systems.

Reliability-centered maintenance (RCM). A systematic process used to determine what must be done to ensure that any physical asset continues to fulfill its intended functions in its present operating context.

Reliability data. A collection of numerical facts based on measuring the motivation of failure by cumulative insults to the component or system.

I don't think anyone is free—one creates one's own prison.

Graham Sutherland

Reliability development/growth testing (RDGT). Type of testing done when the current reliability of equipment or a system must be evaluated, goals for improved reliability established, and test-analysis-and-fix (TAAF) procedure is to be followed until the goals are achieved or exceeded.

Reliability engineering. Appropriate application of engineering disciplines, techniques, skills, and data to access problems or improvements, achieve the required reliability, maintainability, serviceability, exchangeability, availability, and yield of products and processes at a cost that satisfies business needs.

If one advances confidently in the direction of his dreams, and endeavours to live the life which he has imagined, he will meet with a success unexpected in common hours.

Henry David Thoreau

Reliability goal. Quantified reliability level aimed for in early program stages where firm requirements such as formal demonstration tests or acceptance tests are not yet required; used in conjunction with a threshold value to provide guidance for development programs; also used to specify desired organizational achievement indirectly related to, but in support of, product-related goals.

Reliability growth. A positive improvement in a reliability parameter(s) over a period of time due to changes in product design and the manufacturing process.

Reliability indicator. A measure of equipment, process, or facility ability to meet operating availability goals; examples include mean time between failure (MTBF), and overall equipment effectiveness (OEE).

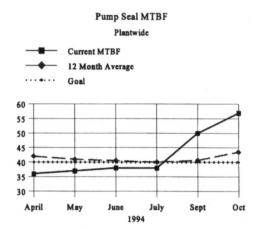

Pump Seal MTBF
Plantwide

■ Current MTBF
◆ 12 Month Average
◆ Goal

It is no use saying "we are doing our best." You have got to succeed in doing what is necessary.

Winston Churchill

Reliability mission. The ability of an item to perform its required functions for the duration of a specified "mission profile."

Every path has its puddle.

Old saying

Reliability objective. Sometimes used interchangeably with reliability goal, but usually applied on a broader scale with less finality as when coordinating subgoals with higher-level organizational goals.

Reliability parameter. A measure of reliability such as mean time between failures, failure rate, probability of survival, or probability of success; reliability-creating factor of the product development process used as an index of design reliability to facilitate trade-offs in the pursuit of reliability achievement or growth.

Reliability qualification testing (RQT). Testing used to verify that a product, which has been built according to particular design specifications, conforms to the intended design reliability standards. Also known as reliability demonstration or designed approval testing.

Reliability target. A general level of reliability typically proposed in the planning or formative stages of a product development program to indicate a desired result when the degree of attainment is still in question.

Reliability threshold. Limit reliability value below which review action is required; used to help control technical and costs risks in advanced development programs.

Relief valve. A specialized valve that is held in the closed position by a spring; it opens automatically when pressure in the system exceeds a preset value and closes again automatically when the over-pressure condition has been corrected.

Never let the future disturb you. You will meet it, if you have to, with the same weapons of reason which today arm you against the present.

Marcus Aurelius

Repairable item. An item that can be restored to perform all of its required functions by corrective maintenance.

Repairability. The ease of restoring service after a failure.

Repair maintenance. See breakdown maintenance and corrective maintenance.

The first problem for all of us, men and women, is not to learn, but to unlearn.

Gloria Steinem

Repair parts. Individual parts or assemblies required for the maintenance or repair of equipment, system, or spares; such repair parts may be repairable or nonrepairable assemblies or one-piece items. Consumable supplies used in maintenance, such as wipe rags, solvent, and lubricants, are not considered repair parts.

Repair time. Actual time used to repair the equipment; wrench time.

Repeatability. (1) A measure of the ability of an instrument to reproduce the same results when subjected to the same test conditions;

sometimes referred to as equipment variation (EV). (2) In alignment, the ability of an instrument to reproduce displayed values when the same input is applied to it consecutively under the same conditions.

Replace. Removal of an entire component and installation of a new or rebuilt equivalent component.

Replacement unit. Any unit that is designed and packaged to be readily removed and replaced in an equipment system without unnecessary calibration or adjustment.

Reprocessing losses. Recycling losses resulting from rejected material or product that must be sent to a prior stage in the process to make it acceptable.

Reproducibility. Variation in measurement readings due solely to the operator; sometimes referred to as appraiser variation (AV).

Request for information (RFI). Means by which buyers canvass a particular segment of industry to determine which manufacturers are interested in, and capable of, providing the component or material in question; request usually is in the form of a business letter stating description of what component, material, or equipment is being sought.

Request for proposal (RFP). A document that describes requirements for a system or product and requests proposals from suppliers.

Request for quotation (RFQ). Document outlining in detail all of the technical and programmatic requirements for equipment, material, software, hardware, or a component as well as any special controls required because of the technical nature of the product.

It is very easy to forgive others their mistakes. It takes more guts and gumption to forgive them for having witnessed your own.

Jessamyn West

Learn what you are, and be such.

Pindar

Reset rate. A measure of automatic reset response, expressed as the number of times per minute that the automatic reset duplicates the correction caused by deviation of the pen.

Resolution. In alignment, the smallest detectable increment of measurement. *Synonyms:* sensitivity, fineness.

Response. A quantitative expression of the output of an instrument (or system) as a function of an input under explicit conditions.

Restoration. Returning the equipment to its original, proper, or ideal conditions.

Resultant costs. Unplanned costs that result from not attaining the required levels of quality and reliability; include internal failures such as scrap, rework, repairs, etc., within the plant, and external failures that are all the costs incurred from failures and malfunctions after the product is delivered to the customer, field services, replacement, warranties, reduced billings, repairs, liabilities, and loss of reputation.

Return on investment (ROI). A financial measure of the relative return from an investment, usually expressed as a percentage of earnings produced by an asset to the amount invested in the asset.

Return on quality (ROQ). The ratio of increase in profit to the cost of quality improvement programs. This measure provides a basis to decide whether or not a quality improvement project is acceptable and to select a better alternative among competing quality improvement programs.

Return on total assets (ROA). A financial performance measure of the relative income-producing value of an asset calculated by dividing the net income by total assets.

There is no greater delight than to be conscious of sincerity on self-examination.

Mencius

Dictionaries are like watches. The worst is better than none at all and even the best cannot be expected to be quite true.

Samuel Johnson

Reverse osmosis. A water purification process where water is forced through a semipermeable membrane under pressure sufficient to overcome osmotic pressure, leaving behind a percentage of dissolved organic, dissolved ionic, and suspended impurities.

Risk. (1) In general, the chance, degree of probability, or possibility of loss, injury, damage, or harm. (2) In a reliability context, the degree of exposure to loss of equipment operational time including the corresponding economic, safety, and environmental impact due to failure. (3) Expressed as a measure by multiplying the probability by the consequence, or the frequency by the severity.

No man ever listened himself out of a job.

Calvin Coolidge

Risk Comparison

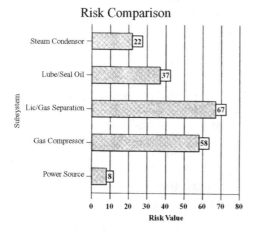

Wrinkles—the service stripes of life.

Anonymous

Risk-based maintenance (RBM). Conducting maintenance based on the economic, safety, and environmental risks associated with equipment failure; prioritizing what equipment to perform maintenance on based on a piece of equipment's position in a weighted risk-rank scheme. The economic, safety, and environmental risks are given a weight factor such that the total weight adds to 1.0. For example, environmental is 0.4, safety is 0.3, and economic is 0.3. The risk consequences of the equipment to be repaired are determined and forced ranked high to low. Maintenance is performed on the high priority items. The list is dynamic and routinely updated as risks change and/or equipment is added.

Risk-centered maintenance (risk-CM). Similar to reliability-centered maintenance using the identical functional descriptions of systems, subsystems, functional failures, and failure modes; however, the criticality class is replaced with an explicit risk calculation, that is, a quantitative value of risk is used instead of a coarse assignment (criticality class).

Risk management. Definite, systematic steps taken to reduce or eliminate the level of exposure or degree of loss; includes risk assessment, trend measurement and analysis, and preventive and predictive maintenance practices.

Every story has three sides to it—yours, mine and the facts.

Foster Meharny Russell

Risk modification. The net change in risk over time from changes in probability and/or changes in consequence resulting from sound risk management practices.

RM. Routine maintenance.

People do not lack strength, they lack will.

Victor Hugo

Robustness. The condition of a product or process design that remains relatively stable with a minimum of variation even though factors that influence operations or usage, such as environment and wear, are constantly changing.

Root categorization analysis (rca). The process of categorizing potential causes for a problem and then creating causal relationships within several categories.

Root cause. The cause that if removed, or prevented from acting, will prevent the primary effect from occurring.

Root cause analysis (RCA). Set of techniques or processes used to identify the causal factors in order to solve problems, whether it be informal or structured in approach; a Total Quality Management (TQM) tool; examples are change analysis, barrier analysis, events and causal factors analysis, tree diagrams, and cause and effect (fishbone) diagram.

Root cause failure analysis (RCFA). Set of techniques or processes used to identify the causal factors for part, component, or equipment failure, whether it be informal or structured in approach.

ROR. Rate of return.

Rotary pump. A positive displacement pump used mainly to pump liquids that are either too viscous or too difficult to pick up suction with a centrifugal pump.

RTBF. Representative time to next failure.

RTF. Run to failure.

S

S. Military (MIL) symbol for 1 failure/10^6 hours.

SAE. Society of Automotive Engineers.

Safety. Actions or steps taken to prevent bodily injury or death. Local, state, and/or federal (e.g., OSHA) laws outline safety requirements for a broad range of activities or situations. Compliance with these regulations often requires specific measures be taken to eliminate or reduce worker's risk of exposure to injury or death.

Fortune favours the bold.

Terence

"Change" is scientific, "progress" is ethical; change is indubitable, whereas progress is a matter of controversy.

Bertrand Russell

Safety factor. Safety calculation for mechanical loading devices, calculated by dividing the average strength by the average stress or load.

Safety valve. A valve on a still, vapor line or other pressure vessel, set so that it will permit the emission of gases when the maximum safe working pressure is reached.

Sample space. Set of all possible outcomes of a random experiment.

Sampling. The act, process, or technique of selecting a suitable sample for chemical analysis or physiological tests.

Scale error. The difference between the true and indicated value of a variable.

Schedule compliance. Ability of maintenance to meet a planned schedule of work, calculated by dividing the actual scheduled hours completed by planned hours scheduled and multiplying by 100. The difference is the schedule breaks due to emergency work. No credit is given for extra or unscheduled work performed beyond that scheduled. *Synonym:* schedule attainment.

A man never tells you anything until you contradict him.

George Bernard Shaw

In my beginning is my end.

T. S. Eliot

Schedule Compliance

Scheduled maintenance. (1) Completed *planned* maintenance that is set to be performed on a specific day or day(s). (2) Preventive maintenance performed at prescribed points in an item's life.

Scheduled preventive maintenance tasks. Number of maintenance work orders generated where work order type is preventive maintenance.

Schematic. Graphic illustration showing principles of construction or operation without accurate mechanical representations.

Schematic design phase. When preliminary drawings/sketches that indicate the scope and relationship of the project components are prepared: includes cost estimates and/or evaluation and initial planning that comprise a concept for the ultimate design of a facility.

The great thing in this world is not so much where we stand, as in what direction we are moving.

Oliver Wendell Holmes

Screening. A process for inspecting items to remove those that are unsatisfactory or those likely to exhibit early failure. Inspection includes visual examination, physical, dimension measurement and functional performance measurement under specified environmental conditions.

The physically fit can enjoy their vices.

Lloyd Percival

SDF. Standard deviation of failures.

SDL. Standard deviation of consumption over LTIME.

Self-directed work team (SDWT). Generally a small, independent, self-organized, and self-controlling group in which members flexibly plan, organize, determine, and manage their duties and actions, as well as perform many other supportive functions. It may work without immediate supervision and can often have authority to select, hire, promote, or discharge its members.

Sensor. Any device for measuring a variable and converting it into a signal that has a fixed relationship to the variable.

Serviceability. (1) Design characteristic that allows the easy and efficient performance of service activities. Service activities include those activities required to keep equipment in operation, such as lubrication, fueling, oiling, and cleaning. (2) A measure of the degree to which servicing of an item will be accomplished within a given time under specified conditions; includes scheduled inspections.

Servicing. The performance of any act needed to keep an item in operating condition (i.e., lubricating, fueling, oiling, cleaning, etc.), but not including preventive maintenance of parts or corrective maintenance tasks.

Set point. The value of the process variable that is to be maintained by control action; the input variable used to control the value of a variable.

Seven management tools. Affinity diagram, the matrix chart, the arrow diagram, the relationship diagram, the matrix data analysis chart, the tree diagram, and the process decision program chart.

Seven quality control tools. Flow charts, Pareto diagrams, run charts, histograms, scatter diagrams, check sheets, and control charts.

SF. Safety factor.

Shear. The strain in or failure of a structural member at a point where the lines of force and resistance are perpendicular to the member.

Shop drawings. Drawings, diagrams, illustrations, schedules, performance charts, brochures, and other data prepared by the contractor, subcontractor, manufacturer, supplier, or distributor

There is only one way to happiness and that is to cease worrying about things which are beyond the power of our will.

Epictetus

Continuous effort—not strength or intelligence—is the key to unlocking our potential.

Liane Cardes

that illustrate how specific portions of the work shall be fabricated and/or installed.

Shutdown. The stopping of equipment or the shutting down of a production line or process unit in order to conduct planned maintenance.

Shutdown losses. Shutdown maintenance loss; time lost when production stops for planned annual shutdown maintenance or periodic servicing.

Total Shutdown/Turnaround Lost Time

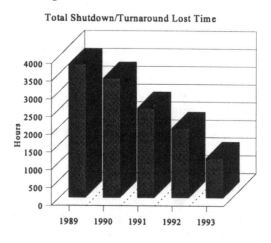

SIC. Standard industry code; scientific inventory control.

Signal. An indication, visual, audible, electrical, etc., used to convey information in an instrument system.

Silo management. Traditional functional form of departmental management in which the organization is managed vertically, often having functional specialists, and not employing an integrative, process-focused approach to product delivery.

Simplification. Principle of reducing the complexity of equipment or systems.

Single point failure. The failure of an item that would result in failure of the system and is not compensated for by redundancy or alternative operational procedure.

SKU. Stock keeping unit.

SL. Service level.

SME. Society of Manufacturing Engineers.

SMRP. Society for Maintenance and Reliability Professionals.

Sneak circuit analysis. A procedure conducted to identify latent paths that cause occurrence of unwanted functions or inhibit desired functions, assuming all components are functioning requirements.

Sneak label or indication. The incorrect labeling of a switch; misleading display.

Sneak paths. A design error that permits the flow of current over an unintended path.

Sneak timing. The occurrence of a circuit function at an improper time.

Speed ratio. Ratio of ideal cycle time to the actual cycle time.

Sole source. The situation in which the supply of a product is available from only one organization.

Sole sourcing. Deliberate decision to use only one vendor to supply material.

Sole-source supplier. The only supplier capable of meeting (usually technical) requirements for an item.

Solid waste. Garbage, refuse, sludge, or other discarded material including solid, liquid,

Mirthfulness is in the mind and you cannot get it out. It is just as good in its place as conscience or veneration.

Henry Ward Beecher

The only menace is inertia.

St. John Perse

semisolid, and contained gaseous matter. Solid waste does not include solid or dissolved material in domestic sewage, irrigation return flows or industrial discharges subject to the Federal Water Pollution Control Act. Special nuclear and byproduct material as defined by the Atomic Energy Act of 1954 are also excluded from the definition.

Solvent. A substance, usually a liquid, capable of dissolving or dispersing another liquid, gas, or solid to form a homogenous mixture.

SOO. Special order only.

SOR. Statement of requirements.

Specification. (1) A clear, complete, and accurate statement of the technical requirements of a material, an item, or a service, and the procedure to determine if the requirements are met; the book of written material that, along with the drawings, constitutes the contract documents describing in full the requirements and terms of the agreement between owner and contractor; sometimes referred to as technical specifications. (2) One of the prescribed limits of control tests used to maintain uniformity of a specified product.

Specific gravity. The ratio of the weight of a volume of a body to the weight of an equal volume of some standard substance. In the case of liquids and solids, the standard is water; in the case of gases, the standard is hydrogen or air.

Spectrophotometry. The use of light for determining color, turbidity, opacity, and other characteristics of fluids.

Speed of response. The time required for an instrument to react to a change in a process variable.

You are remembered for the rules you break.

Douglas MacArthur

Art washes away from the soul the dust of everyday life.

Pablo Picasso

Spill tanks. Holding tanks used to collect chemicals discharged from a chemical storage facility. Usually located away from a facility on a tank farm, individually spill tanks are connected to different waste line systems to prevent the mixing of incompatible chemicals.

SPIN. Specific plant identification number.

SQL. Standard query language.

SR. Stock review.

SRO. Stock reservation order.

Stability. A measure of the change or wander of important parameters of the parts in the lot, or attribute(s) for a product or process over a period of time; a reliability specification; measure used in statistical process control.

Standardization. (1) The process of designing and altering products, parts, processes, and procedures to establish and use standard specifications for them and their components. (2) Reduction of total number of parts and materials used and products, models, or grades produced. (3) The function of bringing a raw ingredient into standard (acceptable) range per the specification before introduction to the main process. (4) The practice of designing equipment or systems using common parts, components, or equipment to facilitate maintenance and reducing the number of maintenance and capital spares.

Standard production rate. In the process industry, equal to design capacity, and expressed quantity per hour or day.

Standby redundancy. That redundancy wherein the alternative means of performing the function is not operating until it is activat-

"When I use a word," Humpty Dumpty said in *rather a scornful tone,* *"it means just what I* choose it to mean—neither *more nor less." "The* question is," said Alice, *"whether you can make* words mean so many *different things." "The* question is," said Humpty *Dumpty, "which is to be* master—that's all."

Lewis Carroll

The foolish and the dead alone never change their opinions.

James Russell Lowell

ed upon failure of the primary means of performing the function.

Standing work order. A work order that remains open to receive and accumulate charges for labor, materials, and/or costs.

Static load. Load that is at rest and exerts downward pressure only.

Statistical process control (SPC). The application of statistical techniques to monitor and adjust an operation; statistical technique to control process parameters by setting naturally occurring control limits.

Statistical quality control (SQC). The application of statistical techniques to control quality of product characteristics rather than process parameters; also includes product acceptance sampling.

Steam. Water in the vapor state.

Step stress test. Form of accelerated test that reveals the uniformity and strength of a product, but does not normally yield failure rate; test repeatedly employs increased stresses according to a prearranged plan.

STKNO. Stock number.

Stockkeeping unit (SKU). An item at a particular geographic location.

Stockless purchasing. Buying material, parts, supplies, etc., for direct use by the departments involved, as opposed to receiving them into stores and subsequently issuing them to the departments, reducing inventory investment.

Storage life. The length of time an item can be stored under specified conditions and still meet specified requirements. *Synonym:* shelf-life.

The two words "information" and "communication" are often used interchangeably, but they signify quite different things. Information is giving out; communication is getting through.

Sydney J. Harris

People exercise an unconscious selection in being influenced.

T. S. Eliot

Strategic plan. The plan to marshal and determine the actions to support the mission, goals, and objectives.

Stress. Force per unit area; when the force is one of compression, it is known as "pressure," an internal force that resists an external force.

Stress to failure (STF). Destructive test where one or more stress factors such as temperature or voltage are increased until an entire sample has failed; the mean value of the STF, and the distribution of failure events about this mean, are used as measures of part reliability of the lot; a reliability specification.

What does not destroy me, makes me strong.

Friedrich Nietzsche

Strip chart. A continuous chart on which a recording instrument places a permanent record, as opposed to a fixed-period circular chart.

Suboptimization. A problem solution that is best from a narrow point of view but not from a higher or overall company point of view.

No, Groucho is not my real name. I am breaking it in for a friend.

Groucho Marx

Subsystem. A combination of sets, groups, etc. that performs an operational function within a system and is a major subdivision of the system.

Success indicator. A measure of an operating parameter that can be trended to identify success in reaching a personal, team, department, or plant goal; examples include routine maintenance expenses, schedule compliance, and maintenance backlog.

Maintenance Costs Per Barrel, Success Indicator

Supply delay time. That element of delay time during which a needed replacement item is being obtained.

Support equipment. Items required to maintain systems in effective operating condition under various environments; includes general and special-purpose vehicles, power units, stands, test equipment, tools, or test benches needed to facilitate or sustain maintenance action, to detect or diagnose malfunctions, and to monitor the operational status of equipment and systems.

SWOT analysis. An analysis of the strengths, weaknesses, opportunities, and threats of and to an organization.

System. A regularly interacting or interdependent group of items forming a unified whole toward the achievement of a goal.

System balancing. The final stage in the installation of HVAC (high velocity air conditioning) systems in which sensors, power controls, dampers, fan speeds, and other systems variables are adjusted to meet the air flow, pressurization, humidity, temperature, cleanliness, and stability specifications for the overall facility.

System effectiveness. A measure of the degree to which an item or system can be expected to achieve a set of specific mission requirements, and which may be expressed as a function of availability, dependability, and capability.

System reliability and maintainability parameter. A measure of reliability and maintainability in which the units of measurement are directly related to operational readiness, mission success, maintenance manpower cost, or logistic support cost.

The will to believe is perhaps the most powerful but certainly the most dangerous human attribute.

John P. Grier

Dost thou love life, then do not squander time, for that's the stuff life is made of.

Benjamin Franklin

T

t. The time variable.

T. Military (MIL) symbol for 0.1 failures/10^6 hours.

TA. Turnaround.

TAGNO. The tag number assigned by the electrical and construction contractor.

TAPPI. Technological Association of the Pulp and Paper Industry.

Temperature. An arbitrary measurement of the amount of molecular energy of a body or the degree of heat possessed by it; a measurement of heat/cold.

Temperature monitoring. Monitoring techniques that look for potential failures that cause a rise in the temperature of the equipment itself (as opposed to the material being processed).

Temporary maintenance. Repair activities that temporarily address breakdowns and repair ailing equipment. The purpose is to get equipment back into operation until routine or corrective maintenance can be performed to effect permanent repair.

Tensile strength. Resistance of a material to a force that tends to pull it apart, usually expressed as the measure of the largest force that can be applied in this way before the material breaks apart.

Testability. A characteristic of an item's design which allows the status (operable, inoperable, or degraded) of that item to be confidently determined in a timely manner.

Nothing which is at all times and in every way agreeable to us can have objective reality. It is of the very nature of the real that it should have sharp corners and rough edges, that is should be resistant, should be itself. Dream-furniture is the only kind on which you never stub your toes or bang your knee.

C. S. Lewis

Space isn't remote at all. It's only an hour's drive away if your car could go straight upwards.

Fred Hoyle

Testing development. A series of tests conducted to disclose deficiencies and to verify the corrective actions that will prevent recurrence in the operational inventory.

Test measurement and diagnostic equipment (TMDE). Any system or device used to evaluate the condition of an item to identify or isolate any actual or potential failures.

Test qualification. A test conducted under specified conditions, by, or on behalf of, the customer, using items representative of the production configuration in order to determine compliance with item design requirements as a basis for production approval. *Synonym:* demonstration.

Fewer things are harder to put up with than the annoyance of a good example.

Mark Twain

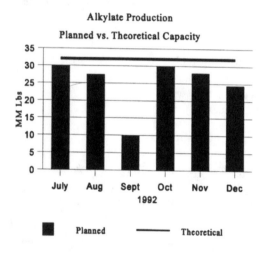

Alkylate Production

Planned vs. Theoretical Capacity

Football combines the two worst features of American life. It is violence punctuated by committee meetings.

George F. Will

Theoretical capacity. The maximum output capability, allowing no adjustments for preventive maintenance, unplanned downtime, shutdown, etc.

Thermal resistance. An index of a material's resistance to heat flow (R); the reciprocal or thermal conductivity (k) or thermal conductance (C). The formula is $R = 1/C$ or $R = 1/k$, or $R = $ thickness in inches/k.

Thermal shock. The stress-producing phenomenon resulting from a sudden, large temperature change.

Thermocouple. (1) The junction of two wires of dissimilar metals that develops an electrical potential that is a function of the temperature. (2) An instrument for measuring temperature by means of the electrical potential produced at a heated junction of two dissimilar metals.

Thermography. Temperature monitoring method checking for temperature variances caused by wear, fatigue, leaks, poor electrical connections, changes in heat transfer characteristics due to delamination of laminated materials, etc. Variations in the surface temperature are detected by an infrared camera and are seen as light and dark areas.

A timid question will always receive a confident answer.

Lord Darling

Thermostat. An apparatus for maintaining and keeping constant any practicable temperature.

It is the mark of a truly intelligent person to be moved by statistics.

George Bernard Shaw

Thermowell. A cavity within a vessel or line, but sealed off from it, for the purpose of inserting a thermocouple or thermometer for temperature measurement.

Threats. Those things or conditions that have the potential to cause harm or an adverse outcome. It is important to identify threats to equipment reliability in order to take preventive or proactive steps.

Throttling control. A type of control that can position the final control element at any position between maximum and minimum limits.

TI. Temperature indicator.

Time. The universal measure of duration.

Time-based maintenance (TBM). Maintenance consisting of periodic inspection, service, and cleaning equipment and replacing parts to prevent sudden failures and process problems.

Time delay. The time required for a specific current or voltage to travel through a circuit.

Time to failure (TTF). Destructive test performed on a small sample of the lot of component parts that are stressed at a level and for a time long enough for the entire sample to fail; the average or mean of the times to failure is considered the TTF for the lot; a reliability specification.

Tolerance. An allowable variation from a specified limit; an allowable departure from a nominal value established by design engineers that is deemed acceptable for the functioning of the product or service over its life cycle.

Tolerance limits. (1) The upper and lower extreme value permitted by the tolerance. (2) In work measurement, the limits between which a specified operation time value or other work unit will be expected to vary.

TOP. Total organizational productivity.

Total asset utilization. Sales divided by total assets.

Total available time. Time equipment could run during a shift or other time interval assuming that there is no downtime, either planned or unplanned.

Total employee involvement (TEI). An empowerment program in which employees are invited to participate in actions and decision making that were traditionally reserved for management.

If a conceptual distinction is to be made, the machinery for making it ought to show itself in language. If a distinction cannot be made in language, it cannot be made conceptually.

N. R. Hanson

I keep six honest serving men. (They taught me all I know.) Their names are What and Why and When and How and Where and Who.

Rudyard Kipling

Total process capability. Calculated by dividing the maximum allowable range of product characteristics.

Total productive maintenance (TPM). A plant improvement methodology that enables continuous and rapid improvement of the manufacturing process through the use of employee involvement, employee empowerment, and closed-loop measurements of results. ("Closed Loop" means the key performance measures or indicators provide feedback to the process to support continuous improvement.) TPM has an operator-performed focus with the involvement of all qualified employees in all maintenance activities.

Teaching is not a lost art, but the regard for it is a lost tradition.

Jacques Barzun

Total quality control (TQC). The process of creating and producing the total composite product and service characteristics by marketing, engineering, manufacturing, purchasing, etc., through which the product and service will meet the expectations of customers.

Total quality engineering (TQE). The discipline of designing quality into the product and manufacturing processes by understanding the needs of the customer and performance capabilities of the equipment.

In certain trying circumstances, urgent circumstances, desperate circumstances, profanity furnishes a relief denied even by prayer.

Mark Twain

Total quality management (TQM). A management approach to long-term success through customer satisfaction. It is based on the participation of all members of an organization in improving processes, products, services, and the culture they work in.

Total quality management tools. The seven quality tools, seven management tools, benchmarking, quality function deployment, and design of experiments.

Transducer. A device that receives energy or a signal from one system and transmits it to another device, where the input and output energies or signals may be similar or different in form.

Transistor. An electronic device using a semi-conducting material, for rectification or amplification of a signal.

Transmitter. A device that modulates a separate power source (pneumatic or electric), in accord with changes in a measured variable and transmits a corresponding signal over some distance.

Trap. A device or piece of equipment for separating one phase from another, as liquid from a gas or water from steam.

Tree diagram. A management technique used to analyze a situation in increasing detail. The full range of tasks to be accomplished to achieve a primary goal and supporting subgoal may be illustrated.

Trend. A relationship between data values defined by three properties (a) trend existence probability, (b) trend type, and (c) trend strength.

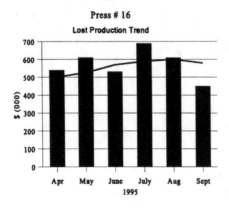

Press # 16
Lost Production Trend

Trend analysis. Critical examination of trend data, focusing on the long-term directional component. Trend defines a long-term movement or direction and serves as an indication of the future based on historical data.

A loud voice cannot compete with clear voice, even if it's a whisper.

Barry Neil Kaufman

A great many people mistake opinions for thoughts.

Herbert V. Prochnow

Trial. A single performance of an experiment.

Tribology. Science of nondestructive testing (NDT).

Turbidity. The degree of cloudiness of water or other liquid caused by the presence of suspended or dissolved particulate or colloidal material. In a spectrophotometric method of analysis, turbidity acts as an analyte (reaction component or chemical) by reducing the transmission of light through the liquid.

Turnaround (TA). Planned shutdown of equipment, production line, or process unit to clean, change catalyst, and make repairs, etc. after a normal run. Duration is usually in days or weeks. It is the elapsed time between unit shutdown and putting the unit on-stream/on-line again.

I have a theory that the only original things we ever do are mistakes.

Billy Joel

Turnaround time. That element of maintenance time needed to replenish consumables and check out an item for recommitment.

Everyone wishes they'd known everything sooner.

Nelson De Mille

U

u Chart. A control chart for evaluating the stability of a process in terms of the average count of events of a given classification per unit occurring in a sample.

Ultrafiltration. A water purification process in which water flows tangentially across a semipermeable membrane having a highly asymmetric pore structure.

Ultra-low particulate air filter (ULPA). An air filter medium used in cleanroom applications to filter out particulate sizes as small as 0.125 micron at 9.9995% efficiency.

Undetected failure. A potential failure identified in the failure mode and effects analysis (FMEA) where no failure detection method is evident to the operator to make him/her aware of the failure.

Uninterrupted power supply (UPS). An electrical power system for a building or an area within a building in which incoming AC power is converted to DC power, stored in batteries, and then converted back to AC electrical to supply building equipment. In the event of an external power failure, power is provided without interruption for a specified length of time from storage batteries. For computer installations, the UPS system performs the additional function of filtering incoming power through the battery packs.

Vagueness and procrastination are ever a comfort to the frail in spirit.

John Updike

Progress might have been all right once, but it has gone on too long.

Ogden Nash

Unit. Collection of equipment, systems, or plant items that are interdependent upon and adjacent to each other; an assembly of any combination of parts, subassemblies, and assemblies mounted together, normally capable of independent operation in a variety of situations.

Unplanned downtime. Time equipment is down due to unplanned events such as breakdowns, setups, adjustments, and other documented stoppages.

Unscheduled maintenance. Corrective maintenance required by item conditions.

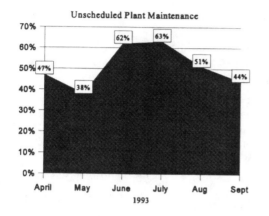

Unscheduled Plant Maintenance

The mistakes are there, waiting to be made.

Sergei Tartakower

Uptime. That element of active time during which an item is in condition to perform its required functions.

Uptime ratio. A composite measure of operational availability and dependability that includes the combined effects of item design, installation, quality, environment, operation, maintenance, repair and logistic support; the quotient of uptime divided by uptime plus downtime.

In an environment of constant change complacency is fatal.

Robert C. Camp

Useful life. The number of life units from manufacture to when the item has an unrepairable failure or unacceptable failure rate.

Utilization rate. (1) The planned or actual number of life units expended or missions attempted during a stated interval of calendar time. (2) A measure of equipment or unit on-stream time expressed as the ratio of actual operating time over planned operating time; often expressed as a percentage.

V

V. The weibull characteristic hours to failure.

Value added. (1) The difference between the sales revenue and the cost of resources (materials and labor) used to produce a product. (2) The actual increase of utility from the viewpoint of the customer as a part is transformed from raw material to finished inventory; the contribution made by an operation or a plant to the final usefulness and value of a product as seen by the customer.

Value engineering and/or analysis. A disciplined approach to the elimination of waste from products or processes through an investigative process that focuses on the functions to be performed and whether such functions add value.

Valve. A device or apparatus used to control the flow or supply of gases, fluids, or fluidized solids.

Valve positioner. An auxiliary air system on air-operated control valves that applies full air pressure on the motor diaphragm or exhausts air from it until the valve stem is positioned as requested by the controller output pressure.

Vapor. Gaseous substance that can be at least partly condensed by moderate cooling or compression.

Vapor barrier. A layer of low-permeance material that prevents condensation within building sections.

Vapor phase. A substance in the gaseous state, under conditions in which it is capable of being liquefied either by pressure or cooling alone.

I know nothing except the fact of my ignorance.

Socrates

Everything not forbidden is compulsory.

The Once and Future King,
T. C. White

Var. Variance.

Variable. A physical quantity that is not constant but varies with time.

Variable cost. An operation cost that varies directly with a change of one unit in the production volume, for example, direct materials consumed, sales commissions.

Variable frequency drive. The process of controlling the speed of fans through motors and controls that vary fan speed by varying the frequency of electrical power delivered to the fan motor.

To persevere is always a reflection of the state of one's inner life, one's philosophy and one's perspective.

David Guterson

Variable pressure. The pressure exerted by the vapors released from any material at a given temperature when enclosed in an airtight container.

Variance. (1) The difference between the expected (budgeted or planned) value and the actual. (2) In statistics, a measure of dispersion of data.

Action translates intentions into reality.

Peter Block

Variance analysis. A comparison between the standard or budget and actual performance; examples include labor and material.

Vendor stocking program (VSP). Raw materials, components, spare parts, and other inventories stocked by the manufacturer or supplier specifically designated for a customer; supplying the items on-demand or through routine deliveries when customers stocks are low.

Venturi tube. A tube, inserted in a line, whose internal surface consists of two truncated cones connected at the small ends by a short cylinder (the throat). As the velocity of the flow of the fluid increases in the throat, the pressure decreases. The tube is used to measure the quantity of fluid flowing or by joining a branch tube at the throat to produce suction.

VESO. A criticality classification system.

Viscosity. The measure of the internal friction or the resistance to flow of a liquid.

VOCAB. A form of materials stock number.

VOH. Value on hand.

Volatility. The extent to which liquids vaporize; the relative tendency to vaporize.

VOO. Value on order.

If you don't know where you're going then it doesn't matter which path you take.

The Cheshire Cat

W

Waste lines. The plumbing that carries liquid waste from the area to centralized processing station. They are usually kept separate from sewage lines in the building to allow special waste processing before the waste is injected into public sewer systems.

Where there is no gain, the loss is obvious.

Chinese proverb

Water heat pump system. Heat pumps are refrigeration machines with a high-pressure side and a low-pressure side. The change in pressure enables the refrigerant to transfer heat by condensing and evaporating at different temperatures.

Wear. The attrition or rubbing away of the surface of a material as the result of mechanical action.

Wearout. The process that results in an increase of the failure rate or probability of failure with increasing number of life units.

Wearout stage. The final stage of failure rate with a sharp rise in the failure rate caused by exhaustion of the component's, equipment's, system's, or product's durability. As components begin to fatigue or wear out, one begins to observe failures at increasing rates for a specified interval.

Weep hole. A small hole in an orifice plate to prevent liquid from accumulating upstream.

Weibull plot. A reliability prediction technique used to evaluate the reliability parameters of components (e.g., electromechanical) and special components (e.g., photoreceptors). It is valuable during the development phase of a component.

What-if analysis. The process of evaluating alternate strategies by answering the consequences of changes to forecasts, manufacturing plans, inventory levels, etc.

White noise. Noise that contains all sound frequencies; it is often used for sound masking in offices and work areas.

WIN. Work identification code.

Work groups. The organization of workers functionally and/or administratively as a unit with a unique and definable mission.

Working capital. Current assets less current liabilities; a measure of liquidity.

Working time. Actual number of hours equipment or the plant is expected to operate; calculated by subtracting lost time from production adjustments or periodic servicing from the calendar time.

Work order. Paper or electronic document specifying the repair needed on a piece of

Senior managers whose strengths lie in their technical knowledge alone are ill equipped to be active players—let alone winners—in the arena.

Art McNeil

I think somehow we learn who we really are and then live with that decision.

Eleanor Roosevelt

equipment. Typically a "needed by date" or priority code (e.g., "E" = emergency, "1" = complete next day) is assigned to each work order to indicate its importance, name, and number for the equipment, location, initiator, and other information. *Synonym:* work request.

World Class Maintenance® (WCM). Process of identifying best practices and comparing an organization's performance level in the practice areas with other companies. Nine specific criteria are used to benchmark organizational strengths and identify improvement opportunities with the objective of increasing competitive advantage through improved system and equipment reliability.

Success is not the result of spontaneous combustion. You must set yourself on fire.

Reggie Leach

World Class Maintenance

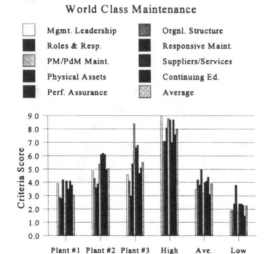

☐ Mgmt. Leadership		▨ Orgnl. Structure	
■ Roles & Resp.		■ Responsive Maint.	
▨ PM/PdM Maint.		■ Suppliers/Services	
■ Physical Assets		■ Continuing Ed.	
■ Perf. Assurance		▨ Average	

Seek first to understand . . . then to be understood.

Stephen R. Covey

WMR. Warehouse materials requisition.

X

X-axis. (1) In a plane Cartesian coordinate system, the axis, usually horizontal, along which the abscissa is measured and from which the ordinate is measured. (2) In a three-dimensional Cartesian coordinate system, the axis along which the values of x are measured and at which y and z are equal to zero.

X bar chart. Chart of the sample averages.

XO. Extraordinary maintenance.

XYZ. The Pareto classification of value on hand.

Y

Y-axis. (1) In a plane Cartesian coordinate system, the axis, usually vertical, along which the ordinate is measured and from which the abscissa is measured. (2) In a three-dimensional Cartesian coordinate system, the axis along which the values of y are measured and at which x and z are equal to zero.

Every definition is dangerous.

Erasmus

YTD. Year to date.

Z

Z-axis. In a three-dimensional Cartesian coordinate system, the axis, perpendicular to the x and y axes, along which the values of z are measured and at which x and y are equal to zero.

Zero-based budgeting. Development of the entire budget from the ground up with no historical data, but rather using the coming year plans as the basis for the budget.

Bibliography

1. Bigliardi, Louis A., *Glossary of Refining Terms,* Detroit, MI: unpublished, 1990.

2. Brigham, Eugene F. and Gapenski, Louis C., *Financial Management: Theory and Practice,* Orlando, FL: The Dryden Press, 1988.

3. Camp, Robert C., *Benchmarking: The Search for Industry Best Practices That Lead to Superior Performance,* Milwaukee, WI: ASQC Quality Press, 1989.

4. Cartin, Thomas, J., *Principles and Practices of TQM,* Milwaukee, WI: SQC Quality Press, 1993.

5. Cox, James F., III, Ph.D. Blackstone, John H., Jr., Ph.D., and Spencer, Michael, S., Ph.D, (Eds.), *APICS Dictionary,* 8th ed., Falls Church, VA: APICS, 1995.

6. Dovich, Robert and Wortman, Bill, *Certified Reliability Engineer Primer,* West Terre Haute: Quality Council of Indiana, 1995.

7. Fitzhenry, Robert I. (Ed.), *The Harper Book of Quotations,* 3rd ed., New York: HarperCollins, 1993.

8. Fogarty, Donald W., Hoffmann, Thomas R., and Stonebaker, Peter W., *Production and Operations Management,* Cincinnati, OH: South-Western Publishing Co., 1989.

9. Forbes Thoughts on the Business of Life, Screen Saver Collection for WINDOWS™, Hunter Communications (screen saver software program).

10. Gano, Dean L., *Incident Investigation and Problem Solving Techniques,* Richland, WA: Appollonian Publications, 1993.

11. Hutchins, Greg, ISO 9000, *A Comprehensive Guide to Registration, Audit Guidelines, and Successful Certification,* Essex Junction, VT: Oliver Wight Publications, Inc., 1993.

12. Ireson, William Grant, *Handbook of Reliability Engineering and Management,* New York: McGraw-Hill, 1988.

13. Hammer, Michael and Champy, James, *Reengineering the Corporation,* New York: HarperCollins, 1993.

14. Jones, Richard B., *Risk-Based Management: a Reliability Centered Approach,* Houston, TX: Gulf Publishing, 1995.

15. Krishnamoorthi, K. S., *Reliability Methods for Engineers,* Milwaukee, WI: ASQC Quality Press, 1992.

16. MIL-STD-721C, "Definitions of Terms for Reliability and Maintainability," 1981.

17. Moubray, John, *Reliability-Centered Maintenance,* New York: Industrial Press, 1992.

18. Nakajima, Seiichi, *Introduction to TPM,* Cambridge, MA: Productivity Press, 1988.

19. Nakajima, Seiichi, *TPM Development Program,* Cambridge, MA: Productivity Press, 1988.

20. Robinson, Charles J., and Ginder, Andrew Paul, *Implementing TPM: the North American Experience,* Portland, OR: Productivity Press, 1995.

21. Shim, Jae K., *The Vest-Pocket MBA,* Englewood Cliffs, NJ: Prentice-Hall, 1986.

22. Stein, Jess (Ed.), *The Random House College Dictionary, Revised Edition,* New York: Random House, Inc., 1988.

23. Suzuki, Tokutaro (Ed.), *TPM in Process Industries,* Portland, OR: Productivity Press, 1994.

24. Tatikonda, Lakshmi U., and Tatikonda, Rao J., "Measuring and Reporting the Cost of Quality," *Production and Inventory Management Journal,* Second Quarter 1996, Vol. 37, No. 2: 1–7.

25. Westerkamp, Thomas A., *Plant Maintenance Manager's Standard Manual and Guide,* Englewood Cliffs, NJ: Prentice-Hall, 1992.

26. Wilson, Paul F., Dell, Larry D., and Anderson, Gaylord F., *Root Cause Analysis: A Tool for Total Quality Management,* Milwaukee, WI: ASQC Quality Press, 1993.